WAKE
of the
INVERCAULD

Very best wishes
Alastair
Looking forward to seeing your
tree ferns when am next in
Bearsden.
Madelene

Wake
of the
Invercauld

Shipwrecked in the sub-Antarctic:
A great-granddaughter's pilgrimage.
Madelene Ferguson Allen

McGILL-QUEEN'S UNIVERSITY PRESS
MONTREAL & KINGSTON · LONDON · BUFFALO

ISBN 0-7735-1688-3
Legal deposit third quarter 1997
Bibliothèque nationale du Québec.

Design, chapter heading illustrations and maps by C. Humberstone.
Map calligraphy by Kirsten Disse.
Photography by Paddy Ryan and Madelene Ferguson Allen.
Illustrations on pages 37, 70, 117, 205 by Michael Stone.
Typeset in Bembo; artwork by Streamline Graphics, Auckland.
Printed by Colorcraft Ltd, Hong Kong.

Published simultaneously in New Zealand, Australia and Asia
by Exisle Publishing Limited, Auckland, New Zealand.

Canadian Cataloguing in Publication Data

Allen, Madelene, 1942 –
    Wake of the Invercauld : shipwrecked in the sub-Antarctic

Includes bibliographical references.
ISBN 0-7735-1688-3 (McGill-Queen's University Press) –
ISBN 0-908988-02-8 (Exisle Publishing)

    1. Invercauld (Sailing ship).   2. Holding, Robert, 1841-
3. Allen, Madelene, 1942-   4. Shipwrecks–New Zealand
–Auckland Islands.   I. Title.
G530.I69F47 1997   910'.9164'8   C97-900305-9

Dedicated to the memory of

Robert Holding

(1840–1933)

and his shipmates on the

Invercauld of Aberdeen

Wrecked, Auckland Islands,

10 May 1864

ACKNOWLEDGEMENTS

This book was written in a small town in rural Quebec; without the help of many from all corners of the world, it would never have come to be. My heartfelt thanks to all:

Australia: Edward Irvine, Long Jetty; the research librarians of The State Library of Victoria; Angela Weeks, Adelaide.

Canada: my husband, Robin, for his support. Alleyne Attwood, Midland, Ont. for her contributions to the story of her grandfather and for the input of my brothers and sisters who grew up with the story. Sophie Demetalin of Superior Travel; Andre Laurencelle, our patient printer; librarians of Bishop's University, Lennoxville, Quebec and the Willard Green Centre, Sudbury, Ontario. The residents of Chapleau and Shining Tree, Ontario. Rose Vaughn, for permission to use the words of her song *Stone and Sand*.

Scotland: Captain A.A.C. Farquharson, of "Invercauld", Braemar, Scotland; Ronald Mac-Gregor, Elgin; Betty Raven, Aberdeen; Stanley Rothney, Aberdeen, who opened so many doors and did so much research; Sheila M. Spiers, of the Aberdeen & North-East Scotland Family History Society.

England: Eric Piercy, Sutton; John Stratford, Kimbolton; the residents of Kimbolton, Ellington and Ellington Thorpe.

New Zealand: Dave Asher of South Coast Video, Riverton, for the opportunity to visit the islands the second time. George Easton, Auckland, for his gift of L. Clifton's manuscript. Lou Sanson and the Department of Conservation for their encouragement and support and permission to visit the Auckland Islands. Pete McClelland, the "Man from DoC", for his enthusiasm and knowledge and constant willingness to be of help before, during and after the "Auckland Island Invercauld Expedition". The Royal New Zealand Navy for the gift of the aerial photographs. Ken Scadden, Director of the Wellington Maritime Museum, for time, information and the Foreword to this book. Ruth and Lance Shaw, Manapouri, and the crew of the *Evohe*, who got us there and back safely. Paddy Ryan of Greymouth, for his photography, and Duncan Sommerville, Wellington.

United States: Jeanne Leveque, Cape Elizabeth, Maine, for reading the manuscript.

CONTENTS

A note on the text: Robert Holding's narrative is denoted by symbol ⛵ *Handwriting and typewriting on chapter heading pages are Holding's own. All other graphics have been prepared for this book and were not part of Holding's original text.*

Foreword

"I am most entertained by those actions which give me a light into the nature of man."

Daniel Defoe

Girdling the earth in the latitudes of the roaring forties and the furious fifties, the twenty-odd island groups which make up the world's sub-Antarctic islands were among the last lands to be discovered, charted and exploited. The combination of atrocious weather, inadequate charting, icebergs and the irregular pattern of these islands has meant countless vessels never returned from the notorious southern oceans. The Auckland Islands group alone has accounted for at least nine known ships, including the *Invercauld*, and undoubtedly many others.

Sub-Antarctic shipwreck stories may be analysed at many different levels. Told simply, tales of shipwreck, castaways and survival read with the breathlessness of a Boys' Own adventure story (as indeed many were used by Victorian and Edwardian authors, with strong moral overtones emphasizing the virtues of courage and self-sacrifice).

On another level, sub-Antarctic shipwrecks were part of the march of human history and technology, a footnote in the saga of 19th century transportation; the sharing of risks by insurance companies, shipowners, passengers and crew alike with vessels travelling the Great Circle Route.

At a deeper level, castaway stories graphically illustrate the power of the human spirit in extreme adversity. Recurring themes in the narratives include: the last match, contemplations of cannibalism, leadership qualities, unshakeable belief in being rescued, in divine providence and the power of sheer ingenuity, tenacity and determination to survive.

One cannot help but wonder about the ability of soft late-20th century humans to cope in such adverse circumstance, deprived of technological gadgets and reduced to the basic elements of survival. Castaway narratives therefore not only tell us of the suffering of earlier generations of seafarers but also something about ourselves.

Wake of the Invercauld is at one time a saga of shipwreck and an odyssey of personal discovery for the author, skilfully woven together into a compelling story. The fate of the *Invercauld* has until now been among the least well known and understood of all the sub-Antarctic wrecks.

Madelene Allen's research has not only connected her with her long lost family and ancestry but has also brought to light a previously unknown account (at least in this part of the world) of the wreck and its aftermath by Andrew Smith, the mate of the *Invercauld*. To have such a detailed castaway narrative is rare, but to have one by all three survivors is, to my knowledge, unique.

Madelene recounts a tale with a plot worthy of any thriller. Holding's narrative carries echoes of both *Treasure Island* and *Robinson Crusoe*, yet is recounted in a down-to-earth, almost matter-of-fact style. Interwoven with Holding's story are excerpts from Captain Dalgarno, Smith, the observations of other sub-Antarctic visitors and the author's own voyage of discovery.

Wake of the Invercauld is a major contribution to the small but significant body of sub-Antarctic shipwreck literature. Knowing both Madelene and her children, I am confident that Robert Holding's blood has apparently not been diluted by the passing generations. If old Robert is up there looking down I am sure he would be proud both of his descendants and the way his story has been told.

Ken Scadden
Director, Wellington Maritime Museum
New Zealand

INTRODUCTION

"No man and no force can abolish memory."
F.D. Roosevelt

OBITUARY – ROBERT HOLDING

After a short illness there passed away early Sunday morning, January 12, 1933, Robert Holding, pioneer of Chapleau, aged 92 years, 10 months and 5 days. This terminated a career wonderful to those who knew him. The late Mr. Holding was born in Spaldwick, Huntingdonshire. At the age of fourteen years he went to sea. In the year 1864 the ship 'Invercauld' was wrecked on the Auckland Isles. Captain George Dalgarno with a crew of 25 men sailed from Melbourne, April 28, 1864, struck May 10. Six of the crew were drowned and 19 washed ashore. The late Robert Holding with the Captain and chief officer were rescued twelve months and ten days later.

Deceased for many years has been known as a Prospector, his quest for gold starting in the early sixties in Australia. He emigrated to Canada with his family in 1888, living for a time in Toronto and then in Kingston; two years where he worked at his trade as a machinist in Kingston Locomotive Shops. He was a veteran member of Toronto No. 235 L.A. of M. In 1891 he moved to Chapleau where for a time he worked as Machinist in the C.P.R. [Canadian Pacific Railways] shops, later deciding to take up mining he went to Goudrean and on Kisanabie Lake. Kisanabie had two boats carrying passengers.

Conditions then changed his plans and his mining ideas revived again, this time to West Shining Tree, making long treks beyond the railway, where at last he was rewarded for his labors. Of late years he has centered his ideas more in his home town, being greatly interested in the new goldfields in Swazy area as he always maintained gold was to be found near Chapleau.

The late Mr. Holding was also the proprietor of the Commercial Hotel here.

At his great age no task was too great and neighbours have been astounded at his wonderful vitality. He was always glad to know of improvements in our town especially along the lines of education. He was instrumental in the early days in having the first school board meeting called and was the first Secretary of the Chapleau Public School Board.

– Chapleau Post

* * * *

The tombstones cast long shadows as we entered the old graveyard in the twilight. Here in the Northern Ontario wilderness I was within a few yards of the end of a two-year journey which had taken me to England, Scotland, Australia and the sub-Antarctic. The wanderer, my great-grandfather, had found his rest in a small railway town north of Sault Ste Marie. A tiny heart-shaped stone marked his final port.

ROBERT HOLDING
BORN AT SPALDWICK
HUNTINGDONSHIRE, ENGLAND
MARCH 17, 1840, DIED JAN. 22 1933.
THY WILL BE DONE

My mother watched as I knelt and sprinkled the sand, which I had brought from the Auckland Islands, over his grave. Spaldwick, Kimbolton, Ballarat, Castlemaine, Talbot, Port Ross, Callao, Kingston, Toronto, Chapleau, Shining Tree. England, Australia, Auckland Islands, Peru, Canada – the litany of places he had touched circled the globe.

I was here to talk to the few who remembered him. An elderly relative of 94 laid a yellowed sealion's tooth in my hand and asked me if I knew what it was. Indeed I did, for I had seen dead sealions and scattered teeth on the same island where Robert Holding had found his. My great-grandmother had worn it mounted on a gold brooch.

Sitting in cozy parlours in houses on streets that he had walked, the old ones remembered:

"My mother told me we had to be polite to Mr. Holding because he was a nice man."

"He was always off looking for gold."

"When he sold his mine he bought an old run-down hotel and the only one who ever made any money from it was the man who sold him his supplies."

"Everyone knew his story."

"We had to be quiet when we went to his house because he was writing."

Seven years before his death, he had begun to type his story "At the strong request of most of my family" – the tale of an incredible year of death and of survival. *The Wreck of the Invercauld* rolled out of his old Remington.

I first 'met' my great-grandfather in 1984 when, after seven years of searching, I discovered my birth family and was given a copy of his manuscript.[1] It was an interesting document: a social history of mid-19th century England as seen through the eyes of a gamekeeper's son, a vivid sketch of the gold mining era in Australia and a survival epic which to me far exceeded the imagination of Defoe in his tale of Robinson Crusoe.

After the excitement of meeting my birth family I found myself returning again and again to the manuscript. Fascinated not only by the ancestral connection but by the whole saga of the shipwreck, I felt it imperative that his narrative emerge from the pages of a dusty duotang. But was it true? This is the question which plagues a responsible writer using first person documentation. You cannot go back in time, but you can find corroborating evidence both in archives and by personal experience. This book is the result of my search for proof.

It is with trepidation that I undertake to write copy for publication, knowing as I do that my qualifications unfit me for such an undertaking. I would therefore ask the reader to excuse all imperfections and assure them that what I do write will be strictly confined to the truth. I have no ambition to write a book, or abilities to do so, but that my experience may prove of service to warrant my spending the time of my leisure, is the only desire of your humble servant. In doing this I am thankful to say I am still in the vigour of good health, and enjoying the blessings of a good memory. My account will involve going into some matters at length. This will involve the mention of many names and places which will be familiar to most of those who this is expected to reach, and will at least shew that I, at 85 years, still retain most of my facilities.

I am Robert Holding, the son of Amos Holding of Keysoe and Fannie Eliza Donoly of Brampton, whose father was a veteran of the battle of Waterloo. I was born, I believe, at Brampton on the 17th of March in the year 1840. We moved to Ellington before I can remember, where my father acted as gamekeeper for Lord Sandwich, who was then living at Hinchingbrook which was once occupied by Oliver Cromwell. The life of a gamekeeper at no time was a bed of roses and there were many occurrences which caused us much trouble and annoyance. There was much poaching and it can be readily understood that we, as children, had to meet the likes and dislikes of those that way inclined and take our share of the insults intended for father.

Much of what I learned was of later value to me in regard to snaring as will ultimately be of much interest to the reader of "The Wreck of The Invercauld". From the age of seven I was called upon by my father for assistance by helping to make snares and also to help in setting them in the evening and going out in the morning to fetch our catch.

Frequently I had been fetched out of bed in the dead of winter between four and five in the morning, when the ground was white with frosts, to walk to Brampton Wood to lift the rabbits and snares. I have often heard my father say his heart ached many times for my condition.

Upon one occasion there appeared at the Bullshead Hotel a man carrying a pack and representing himself as a smuggler and seafaring man. This brought together several of us lads, including your humble servant, to see one of the wonders of the world. Upon this particular night he entertained us with glowing accounts of sea life and offering to engage a number of us for that purpose, did put some of us through imagined drill. Well, from this my mind became quite upset, leaving a strong desire to see more of the wonders of the world. The result was a determination to run away and it was but a short time ere Charley Ellis, Tom West and myself started off one Sunday afternoon to walk to Bedford. We had to pass by the very doors of my poor old grandmother. We got by without notice but on arriving within three miles of Bedford we noticed that father, accompanied by my Uncle Wolston, was following us. It was but a short time ere I was on my way back to Kimbolton. Needless to say the other two delinquents followed pretty closely.

The full account of Holding's early life is in Appendix One. He concludes his introduction:

It is my opinion that the foregoing is sufficient to show how easily the course of one's life may be changed when young, by events of little report. I am sure that there are bitters as well as sweets in any hasty undertaking of the kind here alluded to. This was proven to me up to the hilt on the 3rd day of December in Gravesend on my way to India. By the recording of the foregoing it is the wish and hope of the writer that there is nothing herein contained of an unpleasant nature or in any way offensive to anyone who may happen to be living or to their offspring or any connections.

There are several of these tantalizing references in his story. He never reveals what happened in Gravesend on the third day of December, nor do we hear anything of India, Mauritius or South America – only the story of the *Invercauld*. The stream of the rest of his life is as silent as the river flowing through Chapleau past the graveyard.

Gold was discovered in the West Shining Tree area (85 miles or 140 km south and west of Chapleau) in August 1911. With the death of my great-grandmother in 1913, Robert Holding, now a man of 73, headed into the wilderness once again to search for his eldorado.[2] The Ontario Mining Records in Sudbury hold the only written records of this part of his life.

Late in May, 1912, the writer was instructed by the Provincial Geologist to proceed to West Shining Tree and continue the examination of that area made during September, 1911. Transportation facilities have improved during the year. The regular train service on the railway has been extended to Ruel, sixty-six miles from Sudbury. Two dams were built in the fall of 1911, on the Opickinimika River in order to deepen its shallow portions. This enables small gasoline boats or pointers to run from Ruel to the north end of Allin Lake, which is 1¹/₂ miles from West Shining Tree Lake. Mr. Thomas Clemow, of Ruel, had two boats on the route during the season, giving a tri-weekly service to West Shining Tree Lake. A wagon road will be built during the coming season into the area from mileage 80 on the Canadian Northern railway. During the year considerable development work has been done. Assessment work was performed on a large number of claims, and a number of the most promising properties and adjacent holdings have been surveyed. In several places, shafts of 20 to 50 feet deep have been sunk, and in other places open cuts have been made.[3]

There follows a list of claims which had been made in that year, ending with:

HOLDING'S CLAIM

In September of this year [1913], Mr. R. Holding, of Chapleau, made a promising discovery of gold west of the south end of MacDonald Lake. Surface work was in progress when the writer left the field.[4]

He continued to work his claim until 1914 when he sold out to a Toronto entrepreneur. There are dimly remembered stories of a court case over the sale. The 1920 report states:

Holding: The Holding property is now controlled by R.I. Henderson of Toronto. The deposit consists of numerous parallel quartz stringers of up to four inches and occasionally one foot in width, in amphibolite or hornblende schist. An inclined shaft, at an angle of about 70° to the southeast, has been sunk on the mineral zone to a depth of approximately 50 feet. At a depth of 30 feet, a 10 foot drift was put in to the southwest and some rich gold samples obtained. Although rich samples have come from some parts of the deposit, nevertheless many other parts were found on assay to contain no gold.

A pump operated by a gasoline engine was used to keep water out of the shaft.

On the adjoining claim to the north Chas. Speed reported that he found gold associated with copper pyrites in transparent quartz veinlets similar to the Holding.[5]

Henderson had incorporated the mine in 1922 with an authorized share capital of $5,000,000. However, the Canadian Mines Register indicates that "Holding Consolidated Gold Mines Ltd went bankrupt in 1933."[6] A sealion's tooth, government mining records, two vacant lots where his hotel and his boarding house (the Commercial Hotel and the "Caruso House") once stood – and a heart-shaped tombstone – are now all that remain.

[1]   *Reunion;* Stoddart, 1992.
[2]   *Report of the Bureau of Mines,* Vol. 22, (1913): p.237.
[3]   *Report of the Bureau of Mines,* Vol. 30, (1921): p.49-50.
[4]   Holding continued mining until he was 79 years of age.
[5]   ibid., p.50.
[6]   *Canadian Mines Register* (1935): p.183.

1

Island.

Auckland. Island

"New Zealand's sub-Antarctic islands are some of the least
human-modified environments anywhere on the globe."

Department of Conservation, *Te Papa Atawhai*

REMNANTS of two prehistoric volcanoes, the Aucklands are an almost unknown hiccup of land in the vast reaches of the Southern Ocean. Some 465 km (290 miles) south of New Zealand, they lie just slightly north of an east-west line around the latitude of Cape Horn. Today, this tiny archipelago of islands is the jewel in the crown of the New Zealand sub-Antarctic program, providing an indispensable resting place and breeding grounds for tens of thousands of birds; they are also home of the rare Hooker's sealion and yellow-eyed penguin. Some of the best preserved examples of sub-Antarctic vegetation grow on the Aucklands, sharing the islands with the world's southernmost forests and many plants which are found nowhere else.

Auckland Island, 43 kilometres long by 24 wide (27 x 15 miles), is the largest oceanic island in the Pacific sub-Antarctic. The western coast is a wall of sheer basaltic cliffs rising over 365m (1200 feet) with only two tiny coves in the extreme north which provide possible access to the top of the island. On the east coast deep fiord-like indentations cut far inland. The northernmost of these is Port Ross (Laurie Harbour), the long finger of which is sheltered from the wild winds by Friday, Dundas, Ocean, Ewing, Rose and Enderby islands. Adams Island (10,117 hectares/24,989 acres) in the extreme south, is separated from the main island by a narrow channel forming Carnley Harbour. Disappointment Island, the plug of an ancient volcano, is the only offshore land on the western side.

Lying as it does in the path of the 'furious fifties', those winds which swept the early sailing ships eastwards, its history is full of tales of shipwreck, death, and survival. Between January 1864 and March 1907 there were eight confirmed wrecks on the islands. How many other ships that sailed into oblivion on the dangerous voyage to Europe and the Americas from Australia, New Zealand or the Orient were wrecked on these shores can never be known.

Mystery surrounds a barrel stave[1] which was found near the hut of the *General Grant* cast-

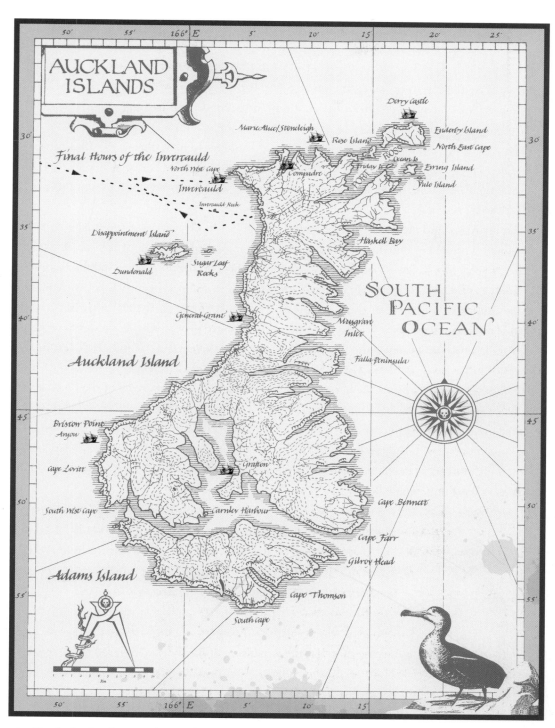

Auckland Islands, showing shipwreck sites.

aways on Rose Island. It was inscribed ' "Minerva of Leith", May, 1865'. Over the years there has been speculation that this was yet another wreck; however as the *Invercauld* survivors had been camped nearby, it is also possible that it was they who added the date to the relic.

The mystery of the disappearances of the *Marie Alice* and the *Stoneleigh* are two more examples. Both were lost with cargoes of wool in 1895, en route from Australia to England. Shortly after the reported losses, the service ship for the castaway depots on the Aucklands found a wreck and scattered bales of wool on North Point. The ensuing enquiry was only able to narrow down the identity to one of these two ships.

The loss of life amongst merchant seamen was shockingly high during the 19th century. The *Encyclopaedia Britannica* records that one in 80 seamen was killed or drowned in 1871. On this one small, isolated island, at least 200 people were wrecked between 1864 and 1907; only 79 lived to be rescued.

The historical uniqueness of the *Invercauld* wreck lies in the fact that the survivors were marooned with no equipment, no food and no shelter at the beginning of the sub-Antarctic winter. In response to the horrific experiences of the survivors of the *Grafton* and the *Invercauld*, castaway depots, containing food, matches, clothing, basic equipment and a boat, were placed on the islands in 1868. These were supplied on a regular basis until 1926 when they were deemed no longer necessary. Each of the other shipwrecked groups was able to retrieve items from their ship, or at some time during their stay avail themselves of materials left at the depots or kill animals deliberately released as food supply for castaways.

Comparisons have been made between the experiences of the survivors of the *Grafton* and the *Invercauld*. It is unfortunate that historians and writers, without greater knowledge of the efforts and initiative of the survivors, have left the impression that the men of the *Invercauld* were incompetents. "...they seemed quite incapable of finding enough food to keep themselves alive. This must have been due largely to ignorance of the natural resources because other castaways have done much better."[2]

Ignoring the fact that the *Grafton's* survivors were relatively well-equipped and those of the *Invercauld* had absolutely nothing, a more recent account reads: "Lacking Musgrave's leadership or Raynal's practical skills the survivors of the *Invercauld* split into two or more groups..."[3] This same phrase occurs in the December 1990 edition of the *New Zealand Geographic*.

Holding's descriptions of the landscape were so detailed, his account so graphic, that I confess to a slight suspicion that perhaps some of it was the expansion of hazily remembered facts. It was a tale of such vividness, such human interaction and such sheer determination to survive that I couldn't resist trying to find out what was contained in the public record.

Accounts of the *Invercauld* are certainly in existence, which makes

	Figure 1: **KNOWN WRECKS OF THE AUCKLAND ISLANDS**			
SHIP	**SITE OF WRECK**	**# ABOARD**	**ASHORE**	**RESCUED**
GRAFTON Jan 1864–Aug 1865	East side of Carnley Harbour – north of Raynal Point	5 crew	5 20 months	3 sailed to New Zealand; 2 rescued later
INVERCAULD May 1864–May 1865	West coast – just south of Column Rocks	25 crew	19 12 months	3 16 died of starvation and exposure
GENERAL GRANT May 1866–Nov 1867	West coast – exact location unknown	25 crew 58 passengers	15 18 months	10 1 died, 4 went for help; never seen again
DERRY CASTLE Mar–July 1867	Enderby Island – Derry Castle Reef	22 crew 1 passenger	8 4 months	8
COMPADRE Mar–June 1891	North Harbour (?) Site now thought to be on north coast. Unknown wreck in North Harbour – perhaps *Stoneleigh* or *Mary Alice*	17	17 3 months	16 1 died
ANJOU Feb–May 1905	West coast – south of Bristow Point	21	21 3 months	21
DUNDONALD Mar–Nov 1907	Disappointment Island – west coast	28	16 9 months	18 1 died

Figure 1: Ingram (*Shipwrecks and New Zealand Disasters*) refers to the *General Grant* survivors finding a barrel stave inscribed: "Minerva of Leith, May 10, 1864 and March 25, 1865." with a large "4" written upon it. The location of this relic and the dates seem to indicate that the dates were carved by the *Invercauld* survivors on a piece of wreckage from the *Minerva* – which would add another ship to the record. The *Marie Alice* and *Stoneleigh* both disappeared in 1865 – wreckage found. No survivors.

it surprising that Ingram in his book *Shipwrecks and New Zealand Disasters* makes no reference to it, despite his extensive research. There were such glaring discrepancies as to the core facts in the accounts which I found that I began to wonder how the errors had crept in and why more details were not known. Some accounts, which had obviously been picked up by successive writers, confused the various wrecks and there are errors not only in the date of the wreck, but the number of survivors, where they lived and the length of time they were on the islands. Barton in *Remarkable Wrecks* wrote: "In four weeks only three were left... They continued to hold out for 12 weeks and 10 days... strange that neither of them (*Grafton* and *Invercauld*) should have met with any signs of the deserted settlement formed in 1850 by the Enderbys..."[4]

The challenge was to find these original sources. Before I read the newspaper accounts, primarily in the *Aberdeen Journal*, the *Aberdeen Herald* and the *Melbourne Post*, I came across the journals of Thomas Musgrave (*Castaway on The Aucklands. The Wreck of the Grafton, from the*

	Age	Home	Port of signing	Fate (death)	Place of death
DALGARNO, Geo. Master	41	Aberdeen, Scot.	Aberdeen	SURVIVED	-
SMITH, Andrew 1st Mate	36	Aberdeen, Scot.	Aberdeen	SURVIVED	-
MAHONEY, James 2nd Mate	?	New York, USA	Melbourne	(Aug 22/64)	Hardwicke
BONNAR, Wm. Boatswain	27	Germany	London	last seen end of May	?
HENDERSON, Alex Carpenter	29	Aberdeen, Scot.	London	(Aug 15/64)	near Hardwicke
PENEST, Richard Cook	32	Gibraltar	London	(May 15/64)	on hills
BARNES, Alex Steward	38	Ayr, Scot.	Melbourne	(July 7/64)	Hardwicke
BORAX, Angel AB	?	?	Melbourne	drowned	west coast
CORRAN, Wm. AB	24	Dumfries, Scot.	London	last seen end of May	?
GOBLE, Wm. AB (Gogland)	36	Rovelden, Scot.	London	drowned	west coast
HANSEN, Fred AB (Fritz)	?	?	Melbourne	(June 10 X) (mid-Aug.)	Lindley Pt.
HIPWELL, Wm. AB	?	Cardiff, Wales	Melbourne	last seen June 4/64	?
HORREY, Wm. AB (Harvey)	?	Yarmouth, Eng.	Melbourne	(Aug 15/64)	Hardwicke

Figure 2: **CREW OF THE *INVERCAULD* of *ABERDEEN***

private journals of Thomas Musgrave) and a translation of F.E. Raynal's book *Shipwrecked on an Antarctic Reef* written in 1866, both of which made reference to the *Invercauld*.

The *Grafton* is certainly one of the best-known of the Auckland Islands wrecks. Captain Musgrave, accompanied by his partner Raynal and three crewmen, had set out from Sydney late in 1863 in search of tin on Campbell Island. Meeting with no success, they attempted to cover the expenses of the expedition by sealing in the Aucklands. On New Year's Eve, they dropped anchor in Carnley Harbour; three days later they were struck by "a gale of unimag-

Figure 2 (cont'd): **CREW OF THE *INVERCAULD of ABERDEEN***

	Age	Home	Port of signing	Fate (death)	Place of death
HOLDING, Rbt. AB	23	Kimbolton, Eng.	Melbourne	SURVIVED	-
GENSON, Rbt. AB	?	?	Melbourne	last seen end of May	?
LAGOS, Ivan AB	22	France	London	last seen end of May	?
MIDDLETON, Wm. AB	17	Aberdeen, Scot.	Aberdeen	drowned	west coast
PAGE, Thomas AB	?	London, Eng.	Melbourne	drowned	west coast
PETERSEN, John AB Big Dutch Peter	25	Sweden	London	last seen end of May	?
SUTHERLAND, Jas. AB	?	?	Melbourne	drowned	west coast
TAIT, John AB	?	London, Eng.	Melbourne	(May 16/64)	west coast
TURNER, Tom AB	?	London, Eng.	Melbourne	last seen June 4/64	on hills
LANSFIELD, Jas. Apprentice	17	Balmoral, Scot.	Aberdeen	(June 20/64)	Hardwicke
LIDDLE, Geo. Apprentice	17	Aberdeen, Scot.	Aberdeen	(July 27/64? More likely June 27)	Hardwicke
WILSON, John Apprentice	15	Balmoral, Scot.	Aberdeen	drowned	west coast
*(*AB - Able Seaman)*					

inable fury" causing the anchor to drag until the ship was blown ashore. Although marooned, they had everything that was on the ship, including the scant provisions remaining as well as firearms and ammunition. With material salvaged from their ship they made a relatively comfortable home while they waited for rescue. Musgrave had taken the precaution of asking friends in Australia to come and search for them after a pre-arranged time. The search was delayed until October 1865, by which time they had built a boat in which Musgrave and two others had sailed to New Zealand[5] for help in a truly courageous and epic voyage.

The story of the loss of the *Invercauld* had reached New Zealand by the time Musgrave arrived and he was naturally astounded to learn that there had been another group of shipwrecked sailors on the island during his sojourn. He told of thinking that they had seen smoke at one time but had discounted it as mist on the hills. Those who speculated that it was strange that the two groups had never met had never been to the Aucklands and could not imagine the ruggedness of the landscape.

Musgrave immediately returned on the *Flying Scud* to rescue his two remaining shipmates. Landing at the site of the abandoned Enderby settlement in Port Ross on the return voyage, they found a body in a ruined building. Thus was born one of the enduring mysteries of the Aucklands: who was this man and how did he die? The only identification was a slate with "some rude hieroglyphics" lying by the body. They supposed that he must have been one of the *Invercauld* crew, but not until the full story of the *Invercauld* was told was this body identified as that of James Mahoney, the Second Mate.

Two groups: one with 100 percent survival rate, the other with a death toll of 22 out of 25. One well-known as a result of two books and countless references; the other, virtually unknown.

[1] Fraser in *South of the Roaring Forties* identifies this object as a barrel. p.112.
[2] Falla, R.A. *A Vanished Township: Hardwicke, or Enderby Settlement*, p.392.
[3] Fraser, C.; p.112.
[4] Barton, G.B. *Remarkable Wrecks*, p.60.
[5] Making landfall on Stewart Island.

2

made sail according

200mile

"Whosoever commands the sea commands the trade; whosoever commands the trade commands the riches of the world, and consequently the world itself."

Sir Walter Raleigh

IT was the winds that determined the trade routes of the world. Ships from Europe bound for the Orient swung westward, almost to Brazil, to pick up fair winds to take them around Africa. Australian wool and grain ships bound for England ran southwards to ride the Westerlies, south of New Zealand and around Cape Horn and into the South Atlantic. North Atlantic sailors dipped southwards on the westerly voyage to catch the northeast trades and headed northwards on the homeward voyage on the westerlies.

The great gold rushes of California, Yukon, Australia and New Zealand sparked a boom in boatbuilding. Ship owners took advantage of this by selling their old ships to companies engaged in transporting the thousands of men to far points of the globe, and placing orders for new, faster ships for more lucrative trade. It was during this last great boom in shipbuilding that the *Invercauld* was built. She incorporated the very latest technologies of wire rigging and steel masts.

During the last half of the 19th century the competing technologies of steam- and sail-driven ships reached epic proportions. As competition for these valuable routes increased, speed became the chief concern of ship designers. The sailing ship was still limited by the direction of the wind; steam-driven ships were handicapped by the cargo space lost to heavy machinery and fuel bunkers, as well as the time lost to put into a port to refuel. The challenge for sailing ship designers was to harness the winds more effectively and the size and speed of sailing ships increased, until it reached its apex with the launch of the seven-masted, 385 foot (117m) schooner *Thomas W. Lawson*, weighing more than 5000 tons gross.

The clipper bow led to the evolution of the famous clipper ships and new sail and rigging designs cut days off the traditional routes. Wire ropes were introduced in the 1840s and by the 1860s they were being widely used. In his *History of Seamanship*, Phillips-Brit explains how it was that rope made from natural fibers was "a fundamental weakness of sail propulsion.

Hemp rope stretches considerably and it was unable to hold the rig steady. The movement of the masts, as shrouds stretched, would cause the slackening of the stays carrying the fore-and-aft sails, which in turn could not then be trimmed for the most effective performance with the wind far ahead of the beam."[1]

Accounts exist of unscrupulous captains who locked the sheets of wire rigged ships so that the sails could not be slackened off by jittery seamen.

Records from Lloyds of London had been easy enough to obtain from the other side of the Atlantic, but to find the type of information gleaned by "letting your fingers do the walking" through the indexes of old books and ancient newspapers only to be found in British libraries, was difficult from Canada. I wrote a letter to the editor of the *Aberdeen Press and Journal* asking readers for information about the *Invercauld*, her skipper George Dalgarno, or her owners Connon and Company.

Within days, three letters with British stamps were in my mailbox. The first was from an elderly lady in Aberdeen who said she didn't know much beyond the fact that a relative of an aunt had been a sea captain who had been wrecked on a "desert island". She said that she had in her possession a photograph signed by one Geo. Dalgarno and would be happy to let me have it.

A Mr MacGregor of Elgin, Morayshire sent a substantial package of information, including a copy of a monograph written by the mate, Andrew Smith. This was an important document, for it is the only other account written by a survivor other than the Captain's report, and it corroborated Holding's record (see Appendix Three). He also directed me to several other sources. When I telephoned to thank him I inquired as to his interest in the *Invercauld*, for his information was so detailed that I was sure that he must have more than a casual interest, but no... he simply had time on his hands and wanted to help an inquiry from so far away.

The third letter was from Stanley Rothney, a retired policeman in Aberdeen, who had "kinsmen" in our small town of 5000 in the Eastern Townships of Quebec. This coincidence led not only to the finding of a stalwart, indispensable assistant but to the development of a friendship.

The spring of 1993 marked the beginning of my travels to prove the veracity of the Holding Manuscript. With my husband Robin, I headed to Cambridgeshire, England. We rooted about in registry offices, talked to local historians then walked through fields to "The Warren", the abandoned gamekeeper's residence near Kimbolton where Holding had lived. Despite the passage of 70 years things were just as he'd described (see Appendix One).

Having a few days free, we decided to take up Stanley's offer of hospitality. The next few days passed in a hectic round of sleuthing and sightseeing in Aberdeen. As we talked we were astonished at the wealth of information which he had gleaned. He had tracked down all sorts of obscure information. We set out on an organized program which included all the sights, the marine museums, the library and an audience with the chief of the Clan Farquharson at his estate "Invercauld." We were all being drawn into the vortex of history.

With the great highland forests nearby, Aberdeen boasted several important shipbuilding yards. Richard Connon & Co 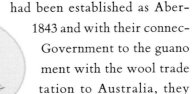 had been established as Aberdeen's first shipbrokers in 1843 and with their connections through the Peruvian Government to the guano trade and heavy involvement with the wool trade and immigrant transportation to Australia, they soon branched into ship ownership. Records exist of their purchase of timber from the Ballochbuie Forest owned by Colonel Farquharson of "Invercauld", for their builder, John Smith. Smith's yard, on Jamieson's Quay just off Market Street, was near where the P&O Shetland Ferry Terminal is today. Known as 'The Inches', this area covered a long finger of land which stretched into the River Dee. It was here that 'Yankee Jack' Smith and his men laid the keel of the *Invercauld* in 1863. Sixty-four shares were issued in 1863: 20 to Connon and Dyer, 16 to John Smith backed by the Commercial Bank of Scotland, 16 to Wm. Gladstone of London, 8 to Grearson and Cole, shipowners of London, and 4 to Robert Middleton, a builder in Aberdeen.

Lloyd's Register of Shipping, 1864/5 describes her:

Rig/Description: Ship, (3 masted, square rigged), felted and yellow-metaled, 1863, copper fastened.
Building Place: Aberdeen
Date: 1863
Builder: Smith
Dimension (in feet): length: 181.7, Breadth: 34.1, Depth: 20.8
Port of Registry: Aberdeen
Flag: British
Owner: Richard Connon & Co.
Master: Dalgarno
Class: +8A1[2]

The *Aberdeen Journal* of 4 November 1863 reported:

There was launched on Wednesday the 28th ultimo, from the building-yard of Mr. John Smith, Inches, one of those fine specimens of naval architecture, which has made Aberdeen so famous for its ship-building for some years past.

The vessel was named by Miss Smith, the builder's daughter, the "Invercauld" in honour of that highly esteemed gentleman, Colonel Farquharson of Invercauld, from whose forests part of the timber used in her construction was procured. She has a figurehead of exquisite workmanship, in full Highland costume, intended to represent the Chief of the clan Farquharson, which is said to be a very striking likeness of the gallant Colonel. The vessel is about 1000 tons builders' measurement, and 900 N.N.M.[3] or thereby; and those who are best judges of these matters say that in her the great essential qualities of a merchant ship – sailing fast and carrying a large cargo – are admirably combined. She will be fitted out with all the newest improvements in hull and rigging, and her saloon and cabins furnished and decorated in a most elegant style. She is to be under the command of Captain Geo. Dalgarno, late of the "Commodore" and is intended for the Australian and Indian trades – is partly owned in London – and is to be under the management of our townsmen, Messrs. Richard Connon & Co. We understand Colonel Farquharson kindly sent a liberal supply of venison and "mountain dew" for the launch. An elegant luncheon, numerously attended, followed the launch.

At 41, George Dalgarno, Aberdeen-born and bred, was at the pinnacle of his career. He had received his master's ticket in 1853 and accompanied by his bride, Helen McMillan, immediately left for Australia, in command of the *Prince Albert*. From 1858 to 1861 he skippered the *Arthur* which ran between Great Britain and Australia. On his return to England he was employed by Richard Connan & Co to command their new ship the *Commodore* and from 1861 to 1863 he sailed to the United States, the West Indies, and Australia. The census of 1861 indicates that after his wife died, his two children Elizabeth and George, who were born in Australia, had returned to Aberdeen and been placed in the care of their aunt and uncle, John and Jane Patterson.

It was during this last posting that he received a telescope from the British Government for "services to the shipwrecked seamen" of the *Hope*, and a silver medal from the United States Government for similar services in the case of the *Franconia*. Now he was to take command of the pride of the Connon fleet, the *Invercauld of Aberdeen*.

With this information at hand, we went to the offices of Cooke's & Sons (successor to Connon & Co) in an old greystone corner building on the riverfront, now alive with ships servicing the North Sea oil rigs. We waited in a tiny room while a clerk searched out records pertinent to the *Invercauld*. The file contained information on two ships – a sailing vessel built in 1874 and a steamship launched in 1912. My inquiry about the 1864 *Invercauld* was met with blank astonishment – they had never heard of an earlier one.

The Fleet List #51, gives a detailed description of the company's holdings:

...Richard Connon & Co. branched out into ship-owning in 1844 when the brigantine *St. Nicholas* was purchased from local owners, and placed on the North American lumber trade. This venture having proved a success, the barque *Mungo Park* was acquired in 1847 but unfortunately her service with the company was to be short, as she was lost off the Farne Islands early in 1850. No additional tonnage was acquired until 1856 when the new ship *Jehu* was purchased from her Canadian owners and builders on her arrival in this country from New Brunswick. In 1859 the ship *Harmonia* took the place of the *St. Nicholas*, the latter having been lost off Aberdeen after her anchor cables had parted in a gale. The build up of the fleet was progressive from 1861 when the *Commodore* was purchased, being followed in 1865 by the *Huano*...[4]

There is no mention in the file of an *Invercauld* until 1874. It was the twelfth ship owned by the company:

(1874-1879) Wooden ship; 223 x 37x 21. 1874: completed in September by Humphrey & Co. Aberdeen. 1879: Sold to Aiken & Co., Aberdeen...Sold to Rob. Mattson, Mariehamn, Russia. 1901: Reported missing in October.[5]

Where is the *Invercauld* of 1863? How could she have been omitted from such a detailed record? Connon & Co was a major enterprise and she was certainly not just an "ordinary ship", but a clipper ship fitted with revolutionary iron rigging – the pride of the fleet, built with the support of the chieftain of the local clan. The loss of a ship with this background, skippered by an Aberdeen captain, and carrying with it six seamen from the local area including the son of one of the owners, would be a tragedy of major proportions. The loss of the ship and subsequent reports of the rescue of three of her crew had been extensively reported in the local papers. A strange business altogether.

* * * *

Leaving the city and dusty records behind, we set off to see the true birthplace of the *Invercauld*. The Invercauld estates border those of Balmoral and in the 1890s the rights to the Ballochbuie Forest were passed "to Her Majesty the Queen, thereby preserving one of the proudest features of the Balmoral amenities."[6]

We drove through the magnificent highland countryside, the first greening of spring on the trees. Passing through modest black wrought iron gates flanked by stone guard posts, we continued for another two kilometers, the River Dee separating us from Balmoral on our left. A herd of deer on the hillside lifted their heads to watch us pass. Rounding the last bend, a scene worthy of a highland film opened out before us – a turreted greystone manor house, with snow-dappled highland hills providing the background.

In the courtyard we were met by a shirt-sleeved individual. A very dirty spaniel, exiled until the mud of the river had dried, was in enthusiastic attendance. We were ushered along a dark hallway to a small room with cutlasses and duelling pieces on the wall, leather-bound books behind glass, ancestors staring down from heavy frames, overstuffed, comfortably dingy furniture huddled round the fireplace. An old Labrador retriever, thankful for anyone who would scratch his ears, kept watch.

We were welcomed by "Mrs. Invercauld", a charming, unpreten-tious woman. Then The Farquharson (as we later learned is the proper reference to a clan chieftain) entered. He apologized for not being in his kilt, yet in tweeds and plus-fours he looked every inch a laird. Although he was most interested in the connection of his clan and estate to the shipbuilding industry in general, and the *Invercauld* in particular, he knew nothing which could shed any further light on the commercial dealings. The information which we needed was probably buried somewhere in the 36 boxes of archives "in the attic" which were unfortunately as yet unclassified.

We were taken to another room where a magnificent wall-size hand-drawn map of the estate hung. It showed a vast area where a chieftain in the 1700s had planted over a million trees for future generations to harvest. A portrait of Cap't Farquharson's great-grandfather, clad in full highland costume, whose likeness on the figurehead had been thrown into the wild sub-Antarctic seas with my great-grandfather, gazed down upon us. Even though no new facts had surfaced we felt privileged to have met the laird, visited Invercauld[7] and seen the remnants of the vast forests that had provided the timber for the *Invercauld*. In memory of her crew, we planted a Scotch pine (which Stanley had nurtured in his garden), by the side of the bordering road.

The publican of a nearby hostelry entertained us during lunch with tidbits of local history. One story from 200 years ago concerned a widow's two sons who had been sentenced to hang for poaching. She had begged and pleaded with the chieftain for the lives of her boys, but to no avail: the sentence was carried out. In grief, despair and anger, she placed two curses on him. The first was that the chieftainship would never pass again from father to son (and it has never done so) and the second was that when the hanging tree fell, so would fall the clan. The tree still stands wired from all the trees around.

* * * *

After our return to Canada, Stanley continued his efforts on our behalf. While searching through Aberdeen cemetery records in St. Clement's Churchyard he found memorial stones to two of the *Invercauld* crew: Alexander Henderson, the carpenter and George Liddle, an apprentice. He seems to have been the only researcher to link the two inscriptions.

IN AFFECTIONATE REMEMBRANCE OF
GEORGE LIDDLE, GRANDSON OF
THOMAS LESLIE, WHO WAS WRECKED
ON THE SHIP INVERCAULD 10 MAY, 1864 ON THE
AUCKLAND ISLANDS AND DIED THROUGH STARVATION
ON 27 JULY THE SAME YEAR WITH 19 OF THE CREW,
AGED 17

The inscription on Henderson's stone is a story in itself of the hardships of those seafaring days:

1866
ERECTED BY
GEORGE AND JOHN HENDERSON
SAILMAKERS, ABERDEEN.
IN MEMORY OF THEIR FATHER
ALEXANDER HENDERSON, CARPENTER,
WHO DIED AT GRAVESEND, LONDON
27TH AUG, 1853, AGED 47 YEARS
ALSO THEIR BROTHER ALEXANDER,
WHO DIED ON THE ISLAND OF AUCKLAND, NEW ZEALAND
15TH AUGUST, 1864, AGED 29 YEARS
AND ROBERT, WHO DIED HERE
ON THE 6TH OF APRIL, 1866, AGED 22 YEARS
ALSO THE SAID JOHN, WHO WAS LOST AT SEA
BY THE WRECK OF THE LOCHIEL OF ABDMABOUT
THE 9TH OF NOV. 1872, AGED 30 YEARS.
*

IN HIS SAD HOME HIS AGED MOTHER
WAITS FOR HER LOV'D SON'S RETURN.
ANGELS OF LIGHT FROM THE DARK WATERS
OF THE FATAL DEEP
HAVE BORNE HIS SPIRIT HOME.
*ALSO THE SAID GEORGE HENDERSON

		Number.	Port of Registry.	Port No. and Date of Register.		Registered Tonnage.	

Invercauld 45,218   *Aberdeen*   23/1863.   888

ACCOUNT OF THE CREW AN[D]

Merchant Seamen's Fund Contributions.					Names of the MASTER and the Crew. Christian and Surnames to be set forth at full length. (see Note 2.)	Age.	Town and County where born.	No. of Fund Ticket (if any) or E 2.	No. of Royal Naval Volunteer's Certificate.	Ship in w[hich] Official Number, natur[e]
Period.		Amount.								
Months.	Days.	£	s.	d.						
					George Dalgaord					Com
					Andrew Smith	36	Aberdeen	"	"	Inver
					William Kourock	27	Russia	"	"	Dapa
					Alexander Henderson	27	Aberdeen	"	"	Com
					James H. Hay	41	D.º	"	"	Kinn
					Richard Reeul	32	Gibralter	"	144w	Com
					William Bonnar	27	Londonderry	"		Govon
					John Marshal	32	Ireland	"	"	
					Maurer Donoral	19	Kensale	"	"	
					Barnard Devlin	33	Ireland	"	"	
					James Reid	35	Lerwick	"	"	
					Charles Quin	20	Dublin	"	"	
					Ivan Lagos	22	Marseilles	"	"	
					Henry Hewitt	28	Ilfracombe	"	"	
					William Coble	31	Hovelden	"	"	
					Christian Larsen	25	Norway	"	"	
					William Corran	24	Dumfries			
					Jacob Thomas Turner	32	Dublin			
					John Petersen	25	Sweden	"	"	
					William Bond	26	London			
					John Power	28	Crosshaven	"	"	
					William Orff	24	Cornwall	"	"	
					David Anderson	28	Norway	"		
					John Wilson	15	Aberdeen			
					William Middleton		D.º			
					John Maloney					
					Robert Holding					

SAILMAKER, WHO DIED AT SEA
19TH MAY, 1880, AGED 40 YEARS.

Stanley wrote: "...[the stone] of the ship's carpenter faces east and is on the opposite side of the church to that of the laddie Liddle. This stone, too, was in a sorry and uncared-for state and almost unreadable because of the green slime and moss. I returned soon after one forenoon armed with hot water, scrubbing brushes and a supply of Ajax. I smartened up the stone which I duly photographed."

The obituary column of the *Aberdeen Journal* of 9 August 1865 lists the deaths of the other two boys:

Lost at sea, in the ship, Invercauld, on the 10th of May, 1864, JOHN WILSON, youngest son of the late James Wilson, Teacher of Music, Aberdeen – deeply regretted.

Drowned at Auckland, 10th May, 1864, from the wreck of the "Invercauld", WILLIAM MIDDLETON, aged nineteen, son of W. Middleton, Balmoral.[8]

They are not forgotten.

Illustration, preceding page: Crew list from Shipping Master's Report – Port of Aberdeen Account of Crew of Foreign Going Ship. Illustration, page 33: graphic based on an original period photograph of Railway Pier, Melbourne.

[1]   Phillips Brit, D.A. *History of Seamanship,* p.273.
[2]   A1 rating for 8 years before next inspection needed.
[3]   New Measure: As of January 1836, the depth and breadth of a ship was measured in six equal parts to give the cross-sectional area to be taxed.
[4]   Somner, G., *Marine News,* 1/65. p.14.
[5]   ibid. p.19.
[6]   *Aberdeen Daily Journal*, 29 January 1901.
[7]   The name Invercauld dates from the 15th century. Inver means "mouth" and the Cala is the small stream nearby. (Cala from the gaelic "a wet meadow".) (Cap't Farquharson)
[8]   The Royal Archives show that William's father was the librarian at Balmoral and that he was responsible for "all the ornamental objects, the library, books, furniture, highland trophies and the like in the castle. He also laid carpets, and arranged the furniture and hangings before he died January 4, 1892."

3

watter smooth

Australia

"Many people's view of the goldfields of last century is that of a romantic period in our history. In fact, the reverse is the case."

Helen Doxford Harris

CAPTAIN Dalgarno had brought the *Invercauld* down from Aberdeen with a shakedown crew including five members of the permanent crew (first mate Andrew Smith and the four apprentices – James Lansfield, William Middleton, George Liddle and John Wilson). Joining the ship in London were two others from Aberdeen: Alexander Henderson, the carpenter, who had earlier served with Dalgarno on the *Commodore*, and James Hoy, the steward.

A mixed lot signed on in London – men from Britain, Ireland, Russia, Gibraltar, France, Spain and Norway. Three of these had served together on the *Conway* and seven others had been shipmates on the *Queen of India*.

Cargo for Melbourne aboard, the *Invercauld* cleared the London docks on Sunday 10 January 1864.

While the *Invercauld* was making fair passage along the west coast of Africa and on to Australia, a 23-year-old Robert Holding, having left the drudgery of England seven years before, was making his way to Melbourne.

I had been six years and four months in Australia less one month to Port Agusta and four months to the Isle of France (Mauritius). In the winters I worked at the diggings and had been occupied during the summer running a steam threshing machine. I was fortunately in good condition, for here as a rule, we were fed on the best.

I had had fair bush experience, having travelled from Two-Fold Bay to the Snowy River diggings, thence through Adelong and on to the Billibung Creek, through to Deneliquin on the Mowina and thence to Bendigo and Castlemaine, Back Creek and Creswicks Creek. I think I can claim a fair knowledge of Victoria. All this may appear out of place here, but it will be seen later that the foregoing was undoubtedly the saving of other lives besides my own.

On the 23rd day of April, 1864 I left Mount Hollowback in the Township of Ascot, seven miles from Ballarat, in the colony of Victoria. My intention then being to go to New Zealand, or to England, being then young and perhaps of a roving disposition. Be that as it may, I had been somewhat upset by something that does not concern the Reader.

I stayed one night in Ballarat and arrived in Melbourne on the 24th, putting up at the Great Western Dining Rooms, which was nearly opposite the Shipping Offices on King William Street. After taking a look up Great Burk Street and feeling lonely in so large a place I began to wish I had stayed up the country. Perhaps I was just too cowardly to return, although a previous employer had told me that he was going to Queensland and that he would be back in three months and would give me the best job he had.

Not having been in Melbourne for over six years I spent the day looking round and took a walk towards the River Yarrow [sic] where there was lots of bustle amongst the shipping. Near the railway station I fell in with three sailors in a position similar to mine. They invited me to join them in a trip to Sandridge. I might state that Sandridge is the principal shipping port in Victoria, and perhaps in Australia. The port is reached by a short train journey of nearly four miles over a sand plain.

On our arrival there was nothing more natural than that we should look the vessels over with a certain amount of curiosity. The large vessels are tied up along both sides of a long pier called the Railway Pier. The first one to take my attention was on the left side and the first man to attract my attention was an old shipmate from some eight years before. John Harold from Sunderland, England, was one of the jolliest fellows it has been my lot to sail with.

They were just clearing up decks preparatory to sailing on the morrow. I called out to him – he looked up and the recognition was mutual although, owing to the lapse of time he scarcely recognized me at first. It was a real surprise to each of us. He came over the side and after a very firm handshake we had a good chat, which was cut all too short owing to the bustle going on cleaning up.

Thus I reluctantly had to part with one of the best shipmates it had been my lot to meet; and whom, I regret to say, I never had the pleasure of meeting again.

* * * *

After a "favourable passage" the *Invercauld* reached Melbourne on Friday 15 April. The call of the goldfields was strong and fully half her crew signed off, including all the men from the *Conway* and *Queen of India*, as well as the second mate and the steward.

Dalgarno, who had left England with a core of men who had been shipmates before the 85-day trip out, was left with the prospect of engaging a number of new men.

The official record shows 11 men signed on in Melbourne. Amongst them was Alexander Barnes, a professional seaman from Ayr in Scotland, who replaced James Hoy as steward. Barnes had had considerable experience on the North American runs and had come out to Australia on the *Jura*. James Mahoney, an American, was engaged as a seaman but was promoted to second mate before the ship left port.

Changes were made in the London crew to reflect the intake for William Middleton received his seaman's ticket "at sea".

We then strolled down the pier and towards the end on the right-hand side we saw a fine looking ship of the Aberdeen Clipper Class – the "Invercauld of Aberdeen". She was a three-master, full-rigged with iron masts, patent reef topsails and wire rigging. Altogether taking to the eye. Hanging in the rigging was a board signifying that they wanted men to sign on the morrow. Someone suggested that we all go and try the ship, and this was quickly agreed upon. She was making her first voyage; everything was new and she looked like making a good ship to sail in every way. She was bound for Callao undoubtedly for guano, (which was the only cargo obtainable). We would pick up the cargo at the Chincha Islands 111 miles from Callao. I had been to these islands when on board the "Emigrant of Aberdeen" some eight years previously. As the next stage of the voyage was to London it suited me admirably.

There were few men in Melbourne at the time so it was an easy matter. We all signed together along with two others of German and Dutch nationality at four pounds per month with one month in advance. It was usual to require a previous discharge but none was required.[1] If there had been I might have been left, being that I had deserted at that very pier some six years and four months previously from the "Emigrant", Captain McLean, Master. A rather strange coincidence.

The Captain of the "Invercauld" was George Dalgarno. A Captain Liddle was in the office with him when we signed on as his son George was on board the "Invercauld" as an apprentice. I well remember Captain Liddle congratulating Captain Dalgarno on the fine looking lot of young men he had engaged.

I spent that night on shore and have every reason to believe that that night I met my brother

William who had gone to Australia to try to locate me. I had not seen him for nearly ten years. Had I known he was out there I might have recognized him and all would have been changed in my future life. I never saw him after as he got married and later died out there. This is one of the regrets of my life.

Our engagement was to go on board on the 27th, which I believe we did to a man. The order now was to prepare for sea and the ship moved down to Williamstown – perhaps to discourage desertion. I am now quite prepared to admit that in my case it might have been a reasonable caution. We were not allowed to go on shore and being about three miles out and no boats available there was little chance of our stealing away. This is where I had been anchored when I first caught sight of Mel- bourne. That was at the time of the Great Hoake. Every ship which sailed from there was packed from deck to main- top with people, many who had closed up shop to search for gold and came back ruined. We soon found plenty of work of the roustabout type to occupy our atten- tion. The ship being wire rigged through- out with patent Top- sails etc. there was little to do to the rigging aloft so we were occu- pied chiefly on deck and in the hold. Our work appeared chiefly to be to find something to occupy our minds so that we could get acquainted with our duties. I must confess that I did not relish the work much after being so long from it.

This continued until the 2nd day of May, 1864, when we left Melbourne in ballast with a crew of twenty-five men, all told, bound for Callao, to take in a cargo there for England. We anchored again just below Gelong nearly opposite Balereen East, which I had visited some few months before and must say would have liked to visit at this time. This was at the entrance to Hobson's Bay, which is very narrow and is one of the finest harbours in the world, about 30 miles in length.

The tide is very strong at the Heads so we had to wait until it was suitable to pass through with safety.

On the 3rd day of May we again lifted anchor, this time to leave one of the most beautiful countries on the face of the earth, where wealth and liberty abound. The day of sailing is always a busy one for sailors and this was no exception. We were on the move early for sailors at that time had no such thing as a six or eight hour day. After clearing the Heads we had a nice light, fair breeze of about five knots which increased gradually as we cleared the land.

By this time we had got the decks in pretty good shape so there was little do. We thus had a little time to ourselves with a certain amount of liberty and could take stock of the land as we passed heading for the Basses [sic] Straits.

I may here be excused when I say I half regretted the step I had taken and wished myself back at Ascot, although I little realized what was in store for me.

[1] "The discharge of a seaman, like his engagement, must take place before a superintendent or an officer of equivalent authority. The seaman is entitled to receive a certificate of service and discharge." Merchant Shipping Act 1854.

4

permit

details

"Hope, like the gleaming taper's light
Adorns and cheers our way"

Oliver Goldsmith

I N December 1990 we began to plan for our first sabbatical in three years' time. We had dreamed of going to Australia and New Zealand and now that our two children had grown, this was to become a reality.

Over dinner one evening I hauled out the atlas and noted the "short" distance between New Zealand and the Auckland Islands. The seed of a mad idea was sown. More out of curiosity than anything else, I phoned the New Zealand Consulate in Ottawa.

"Is it possible to travel to the Auckland Islands?" I asked.

Long Pause.

"Where are they located?"

"About 300 miles south of New Zealand."

Pause.

"Are they ours?"

(Woops! What were we getting ourselves in for? Even a New Zealander didn't know about these islands!)

"You could write to the Hydrographic Office in Auckland."

My first letter was somewhat tenuous as we had not yet actually admitted to ourselves that visiting the Aucklands was a viable venture. Weeks passed while my letter was passed on to the Department of Conservation in Invercargill which passed it on – and apparently on and on, for in very short order I had replies from three very different sources.

On 28 February 1991 I received a letter from Pete McClelland, who would become my support, mentor, a fount of information, and finally our guide and friend. This first letter brought me into a world of history, wrecks and maps. I learned that the *Invercauld* was one of four wrecks on the Auckland Islands which had not had a book written about it.[1]

Pete wrote: "Regarding maps – at present there is only an out of print map which was done totally by land-based surveyors in 1943-44.[2] The new topographical map (approx. 100 x 60 cm) is due for publication within the next 2-3 months. We have, however, been sent a copy of the draft for comment on place names and as such I have photocopied it in parts for you. It is somewhat of a jigsaw but it is the best we've got."[3]

I knelt down on my office rug with the pile of folded papers. It was indeed a jigsaw, but soon what I had only seen as a three-inch blob on an atlas map of the immense South Pacific exploded into a long finger of hatch marks, contour lines, indented coastlines and sea. Sitting cross-legged on the floor, Holding's manuscript on my knees, I reread his story following his route. The amazing thing was that it was possible to follow it.

A few days later a letter arrived from Ken Scadden of the National Archives, who had been doing historical consultation work for the Department of Conservation: "Ironically one of my tasks this year was to endeavour to contact descendants of Robert Holding to ascertain if he had left any papers or journals or photographs. The last address I had for him was 'Box 48, Chapleau, Ontario, Canada.' My search strategy was to have been to search for Holding descendants in or around Chapleau using the telephone book. Would this have led to you, I wonder?"

In August, a large package of material arrived, with a letter from a Duncan Sommerville: "Your letter has been passed on by Survey and Land Information. Maybe I can be of some assistance. I have had an interest in N.Z. offshore islands. A large part of my work has been as a diver and fisherman. My main interest has been sailing ships. I have also been privy to several attempts at recovering the gold of the *General Grant*. The research on this ship has taken me down many tracks but not directly to the *Invercauld*. You have fired my interest. As you may guess, I am very interested in your great-grandfather's journal; perhaps it has answered many questions for no one else speaks for the *Invercauld*."

Duncan's last sentence became a talisman: "I have dived many wrecks and know of many lonely graves and the lasting thoughts I have are always with the lost seamen."

Included were copies of the *New Zealand Pilot*, relating to the Auckland Islands, a photocopy of a slate with part of the word "unknown" left by Mahoney's grave, a history of the Maori on the Aucklands, information about the survival depots, Campbell Island history, pictures from *Yesterday's Gold* of the west coast and articles by R.A. Falla from the Auckland Island expedition of 1972-73. By the time I had worked my way through all this material the somewhat wishful thought of visiting the islands had become a driving desire to get there – somehow.

I don't remember our consciously making the decision to go to the Aucklands. Letters of inquiry and information went back and forth, and like Topsy, the project "just grewed". It was almost a year later that I wrote Pete asking specifically how one would get to the islands – if one should want to go. His letter did not leave us wildly optimistic as to the successful completion of this madcap scheme.

This administration [Department of Conservation, from here on referred to as DoC] includes permitting and overseeing all visits to the islands. As you will understand, being nature reserves, access to the Aucklands is limited and tightly controlled.

Visits fall under two categories: research – which allows staying overnight on the islands (Enderby and Auckland) and should be of benefit to the islands, and tourism – which covers all other visits and doesn't allow overnight stays. All visits to the island must be accompanied by a departmental representative generally at the visitor's expense (around NZ$2000 depending on length of stay) plus providing transport. There is also a permit application fee of NZ$1000 once you have set your itinerary.

All research applications are judged on their merits and, as I presume you wish to stay on the islands, we would need a full breakdown of the work you wish to do on the islands.

Access to the islands is difficult and expensive. Transport can be through either a boat charter – generally around NZ$2000 a day; or getting a berth on a tourist boat which visits the islands. Both are strongly weather dependent (last year a charter boat was held up for seven days by bad weather).

As soon as you get a better idea on the logistics of the trip i.e. transport and the feasibility of being dropped off and picked up where you require, number of days on the island and where you wish to go (as accurately as possible), we'll be in a better position to judge the benefits of the trip to the department and its eligibility for a non tourism entry permit.

By this time, friends and family were beginning to hear of our project and the question Why? was asked over and over again. Now, if we had any hopes of getting a permit, I had to articulate the whys. Certainly, my historical and geographical background and a love for travel did play a part, but more importantly, it was the magnetism of Holding's account. Our strong feeling was that if Robert Holding's descriptions of the landscape proved exact after 55 years, then it was likely that his account of the events of that year was also true. The driving force to correct the historical records and to finally see the places where the men of the *Invercauld* had lived and died, carried us along.

A year later I wrote a mission statement to the Department of Conservation:

9 September 1992

Re: Permission to visit the Auckland Islands
Reason for proposed visit: To locate the site of the wreck of the *Invercauld*, lost 10 May 1864, and to follow the wanderings of the survivors, one of who was my great-grandfather Robert Holding. The outcome of this visit will be a book to bring his memoirs (original 115 pages) to public attention and to link them with the present-day Auckland Islands.

It is my understanding that apart from the official record there is no account of the year

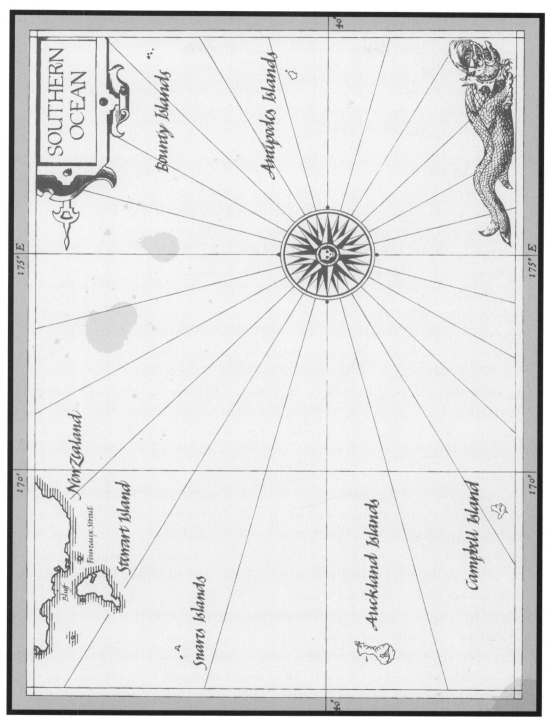

New Zealand's Southern Ocean islands.

spent on the island by the survivors, thus the document in my possession reflects significantly on the understanding of what happened to the 19 original survivors and why there were only 3 survivors rescued a year later. The role played by Robert Holding was nothing short of heroic and I would like to bring this to the attention of the public.

His memoirs (written 1918-1919) contain vivid geographic detail, and from this I would foresee being able to follow very closely their route from the site of the wreck (just south of North West Cape – established by air photos and the journal), across the island to Port Ross and across Rose Island to Enderby Island.

Itinerary:

1. Landfall at Port Ross.
2. Crossing the island directly to establish visually the site of the wreck.
3. Work our way back to Port Ross following their explorations along the north coast.
4. To go to Rose Island and Enderby Island, the site of their rescue in May 1865.[4]
5. If possible, to view the site of the wreck from the sea.

A written and photographic record would be made throughout – linking our visit with his, 127 years ago.

As an offshoot of this story, I am also interested in the possible effect of wire rigging on the outcome of the wreck. The *Invercauld* was a wire-rigged ship which resulted in there being no canvas or ropes to help in survival since everything sank. I would also like to look into the inquiry into the wreck, for the statement of the master, George Dalgarno, is at odds with Robert Holding's records. If possible I would like to try to trace his family and that of the other survivors.

I do hope that this request will be considered positively in the light of both its historical value and personal interest.

Getting to the islands was certainly proving to be a challenge; Pete McClelland had sent a few suggestions of where to begin. A glossy brochure arrived from the tour company with a friendly letter suggesting: "...that you join our cruise departing from Bluff on January 24 and we will transport you to the Auckland Islands and weather permitting take you to the site of the wreck by sea and we will also land you on Enderby and Rose Islands. If you feel you are capable and equipped for the return walk to North West Cape we will put you ashore in Port Ross. We will continue on to Campbell Island returning via the Auckland Islands to pick you up and return you to Bluff on February 3."

Great! But – what would happen if we were delayed in the hills, what would happen if they were delayed? One person in the family was quite enough to be marooned on these islands. Pete's account of the charter boat being delayed for a week was not something to be taken lightly.

Pete dismissed the idea of hitchhiking down on a government vessel: "Regarding government trips for transport – put simply, I would not plan on it. The only work we have currently got planned is a follow-up to the rabbit eradication on Enderby and Rose which will be in about March, and that will probably be via tourist trips. Any other work will be dependent on the navy and it is often at fairly short notice."

The other possibility was to charter our own boat. Further correspondence with another company elicited the following information:

"The daily charter rate on a smaller ship is approximately NZ$3000 per day and you would have to allow at least 10 days, even then depending on weather and sea conditions, it might not be able to get you there; or to charter a bigger ship such as we have which can handle most sea conditions, would cost you in excess of NZ$5000 per day."

Definitely not on – unless grants were forthcoming. So, in the middle of a world-wide recession, I ventured into the dark waters of grant applications. The responses were not unexpected, but certainly disappointing.

I regret to inform you there are no funds available in the current financial year to provide a grant in aid for your project. I hope you have been more successful elsewhere...

Thank you for your letter; the fate of the survivors certainly makes fascinating reading. I am afraid that this university is in no position to sponsor the sort of travel that you have in mind...

Although your project is a most interesting one, we cannot think of any way to provide funding for it...

You are involved in a most interesting and worthwhile project. Unfortunately we do not have any funds which are available to support you. We do have a competitive set of awards for researchers in New Zealand history; but I am afraid they are confined to New Zealand citizens. Nor can I think of any other likely source of support. Sorry about that. Good luck on your project...

And so on. I soon got to the point of not even wanting to open letters with a New Zealand stamp unless they were from DoC or Ken Scadden. It seemed we would have to manage this on our own. If we could find a boat large enough to take other researchers we could offset some of the cost. However, we had applied for a permit for our project; others would have to apply individually and it was not a foregone conclusion that permits for other projects would be granted.

An unexpected complication was that not only do travellers to the Aucklands need permits, but skippers must have sub-Antarctic permits. We wrote to them all. One skipper was trying to sell his boat and could not promise to be available at the time of our trip, another had moved and so it went until we wrote to the owners of the *Evohe*. This sounded most promising for the cost was somewhat more reasonable and she was available. We perused the brochure describing this 87 tonne, 25 meter (82 foot) steel-hulled sailing ketch with joy and excitement. Then we heard she had been leased elsewhere. However, the new skipper Lance Shaw was ready and willing. Yet we were still awaiting the final word about our permit.

A flurry of correspondence across the Pacific ensued. I shall always treasure the following paragraph from Pete at the time we were discussing our goal of walking across the island to the site of the wreck above the western cliffs: "Frequently walking on the tops is dangerous, if not impossible due to weather i.e. very strong winds and rain and worst of all thick clag (low cloud) which makes finding your way nearly impossible even with a compass. In which case all you can do is camp and wait it out. Sorry to sound pessimistic. There is every chance you could get down there to fine weather and have no problems getting everything done. But I believe in planning for the worst and as you will be aware from your great-grandfather's writings the worst that the Aucklands can put out is pretty bad."

I had sent Pete a copy of Holding's manuscript which reached him just before his summer six weeks on the island. The next letter arrived at the end of a long northern hemisphere winter: "Finally, many thanks for the manuscript – I took it south with me and read it while on Adams Island – it makes you realize many of the luxuries we now take for granted, even when 'roughing it' in the outdoors."

Finally, on 3 March 1993, six months after I had sent our mission statement, the word which we had been waiting for arrived by fax: "No problem with the permit. Let us know how your planning is going – good luck with the transport. Pete."

We were both relieved and apprehensive to realize we could go. I faxed the Shaws of the *Evohe* and contracts were soon signed.

The weeks before our departure were hectic. Lance's partner Ruth became a firm friend through her letters. Her communications with DoC were a great help as the numbers from our original permit application grew with the addition of our son Bruce, my brother Jim and a New Zealand photographer, Paddy Ryan.

Early in October 1993, we left the scarlet hills of Quebec behind and headed south-west across the Pacific.

---

[1]   The others being the *Derry Castle*, *Compadre*, and *Anjou*.
[2]   The coastwatchers.
[3]   I worked with this map for the next 18 months.
[4]   We later confirmed that the rescue was from Rose, not Enderby as some accounts state.

5

took in fore-top-galan

land ahead

BEFORE me is a sepia photograph of a sealskin boot hanging on a wall. When Robert Holding emigrated to Canada in 1888, he brought with him the boot which (so the story goes) he had been wearing when rescued (the other he gave to the captain of the ship). This boot and the three remaining matches were among his most prized possessions. In 1926 a boy's band from Melbourne performed in Chapleau, and he gave them the boot with the intention that their museum place it on display. Somehow it has ended up as part of the La Trobe Collection in the State Library of Victoria.

So far from public view is this relic, that it is necessary to make an appointment some time in advance to view it. I appeared at the appointed time, was led into the 'rare collections' room where, lying on the table, was a two foot square, white styrofoam box. Putting on the gloves provided, I lifted the lid and there, encased in a plastic bag, was *the boot*. Attached to the bag was a hand-written note: "sealskin boot, Invercauld, Rbt. Holding. It smells." Reverently, I slid it out and held it. How I wished I had known the old gentleman – to have been able to ask him to tell me its story.

Judging by the size of the boot, Holding had not been a big man. Carefully crafted and in remarkably good condition, it looked very much like something you might find in a wilderness wear section of a modern sports store. The hair on the outside was smooth to the touch, and almost unmarked; the leather was firm, but not brittle. I have an old pair of leather mukluks in my basement which are not in as good condition, and this boot was a 130 years old.

I replaced the boot in its container, took off the gloves and laid them on top. This artifact, the sole remaining item from the *Invercauld* disaster, does not belong in a styrofoam box deep in the bowels of an Australian library, a cardex file the only clue to its existence. New Zealand has the only museum display in the world devoted to the sub-Antarctic, and it is here where this boot should rest. The Southland Museum has written, as I have, to enquire about the possible transfer of this artifact. The reply from the La Trobe Collection librarian was blunt and to the point:

> The descendants of donors of items to public collections often feel that their forebears made mistakes in selecting suitable places to deposit such items, or that they should have remained in the family. This is understandable, and most public institutions including this Library have generous loan policies and procedures which mean that works from their collections can be seen in a wider context. The request for loan of the boot to the Southland Museum will be assessed according to the Library's loan policy and condition reporting by the Conservation Department which will determine whether and how it is to travel. No doubt the Museum will advise you of the results of their request.

With four days remaining before our flight to New Zealand, we left museums behind, rented a car and headed for the goldfield area north of Melbourne.

The flat plain inland from Melbourne would have posed no difficulties for the long trail of men walking northwards, but the land then rises steeply up a long hill. Diaries of the time record that the road was littered with equipment dropped by those who placed speed over need. A sleeping roll, a pot, a pick and a shovel were the only true necessities.

Zipping along the four-lane highway at 70 miles (110 km) an hour, we soon reached the area which had attracted thousands of men from all over the world. Sovereign Hill at Ballarat is a reconstruction of the 'typical' mining town of the day. Pay your entry fee – walk the dirt street peopled by actors in costume, dodge the stagecoach, pan for gold, take a guided tour in a mine with tableaux along the way. It's all there, and yes, well done, for the tourist. But this was not the Australian mining lore for which we were searching.

In the back country it is still possible to find a brick chimney here and there which marks old workings. Narrow paths over a landscape once scoured barren by high-pressure hoses, scrabbled and dug by men with a dream, is now gentled by a century of growth. Rusting pipes, piles of gravel and stream diversions mark the forgotten mining country.

Talbot, as Back Creek is now known (see Appendix Two), heralds itself to the world as 'Home of the Yabbie Festival.' We didn't take time to explore for tiny crayfish; we had come to see Willie Nutt's hotel on Camp Street. It wasn't difficult to find, for Talbot has

Artifact from the La Trobe Collection, Melbourne, Australia: "sealskin boot, Invercauld, Rbt. Holding. It smells."

been forgotten by the world. It is as close to a ghost town as you could find. The tiny population has worked hard to keep its town alive. The museum holds a wealth of records, photographs and relics. Here we found that Willie Nutt had paid his taxes and was buried in Amherst. This once thriving town a few miles from Talbot is now nothing but a cemetery. A few stone markers stand, but most are worn, scorched wooden boards. Willie Nutt's marker had been consumed by one of the numerous forest fires. His one-storey hotel, its pink paint bleached almost white by the unforgiving sun, still stands, windows boarded. As I stood in the street, which was empty but for two sheep grazing beside a house, I felt the presence of the friendly ghost of Grampa Holding.

It was harder to find him on Railway Pier back in Melbourne. We had boarded the one car electric train and followed the route across the undulating sand plain that he and his mates had travelled 130 years before. Disembarking, we walked to Railway Pier.

In place of a line of tall masts, the great ferry *Abel Tasman*, decks lined with passengers, towered above us. A modern building containing shipping offices and waiting rooms arches over the railway line which still runs to the end of the pier. Although this 'new' pier was completed in 1930, the place marked his point of departure.

With great blasts of her horn, the *Abel Tasman* cast off, ropes snaking up into gaping openings. She reversed out into the bay, turned and headed south towards Tasmania. The sky, red with the setting sun, deepened into darkness as we waited for the train back to the present.

* * * *

Melbourne was the southern gateway to Australia and despite the vast numbers of ships arriving and departing, it was an underdeveloped port with no wharf until 1842. Horses were used to tow the smaller ships up the River Yarra. Ships over 150 tons had to anchor in the bay and transfer cargo to lighters. As a consequence, many vessels stopped at Williamstown, a cluster of houses three and a half miles further down the bay near the entrance to the Yarra. *The Melbourne Advertiser* criticized the landings at Williamstown where "people had to wade ashore across mud flats because there was no jetty."[1]

Once the 110 foot (33.5m) bluestone boulder jetty had been built with convict labour, Williamstown became the official port for Melbourne, despite the 10 mile (16 km) trip across the Sandridge plain, and a ferry across the river.

The first immigrant ships from England began arriving in 1839 – "in one week over a thousand passengers arrived in four ships. This surplus of manpower was absorbed by the building of the first carriage-way from the south bank of the river at Melbourne to Sandridge."[2] From 1851, with the discovery of gold in Victoria, tens of thousands arrived from all over the globe and within seven years the population had exploded from 76,000 to almost half a million.

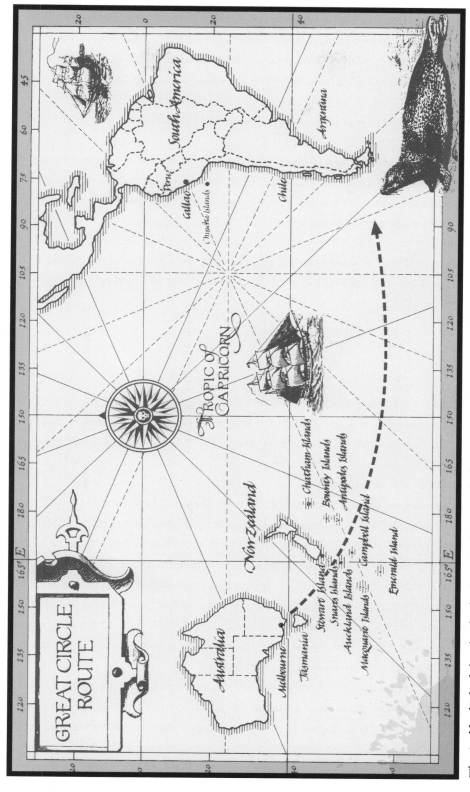

The Auckland Islands lay close by the Great Circle Route, the fastest passage from Australia to Britain.

The surface of Hobson's Bay was covered with sailing ships – "the effect was a forest of masts and yards. It was dense enough to obscure the opposite shore."[3]

The Yarra River was simply too small to accommodate this influx of ships so a new port was developed at Sandridge (also known as Port Melbourne). In 1850 a major jetty, 400 feet long (121m) and 15 feet (4.5m) wide, extending into eight feet (2.4m) of water was completed. The Hobson's Bay Railway to Sandridge opened in 1854 and shortly afterwards Railway Pier was constructed.[4]

History was made at Railway Pier. Thomas Austin, an avid sportsman, missed his rabbit shooting so ordered some sent out from England. He picked up his consignment in 1859. Unfortunately, his skill with a gun did not keep up with the breeding pattern of the rabbits and soon they had run wild, becoming a curse to succeeding generations of Australian farmers.

The largest steamship in the world, the *Great Britain*, plied the seas between London and Melbourne. In 1864, the same year that the *Invercauld* tied up at Railway pier, the Confederate raider *Shenandoah* arrived and much to the disgust of the U.S. consul, her crew received a royal welcome.

When we left the weather was very fine, but during the night it commenced to blow hard, and after a good deal of trouble and fatigue we succeeded in getting safely though Bass's Straits, at all times a very dangerous passage, owing to the great number of sunk [sic] rocks and shoals that beset the passage. The wind still continued to blow very strong, with showers of sleet and snow.

Having reached the open sea the watch was chosen. The Captain's watch is called the Starboard Watch, but as he doesn't take it the Second Mate acts in his stead. If my memory serves me right, he has the first choice or pick of those he thinks likely to be the best men. The Mate has the Port Watch and the second choice and so on alternatively until all have been chosen with the exception of the cook, steward, carpenter and sailmaker etc. who sleep in all night except when all hands are called. Once mustered for the purpose it did not take long to decide the matter and I found myself selected for the Port Watch. At this long time since the events occurred which I am about to relate, to wit nearly 55 years, I could not pretend to state who was in the same watch as myself.

We had now took our regular watch and the other watch went below. There was not much to do as it was too soon to cathead[5] the anchor and stow away the cable so we had a little time to get better acquainted. As it ultimately turned out we never had time enough together to become fully acquainted with some of our shipmates. The wind steadily increased as we travelled Southward. We now got the anchor gateheaded and the cable stowed away.

There was nothing of importance to do aloft, owing to the trim state of the rigging nor to do on deck as the ship was in ballast, (that is ballasted with rocks for steadiness). We thought of the good time we would have as there was no apparent necessity to expect much work for some time. But this couldn't last on board ship – the powers that be have a way of making work whether it is required or not and they felt that we must be doing something to keep us in exercise. We were not supposed to be the judges of that.

The order went around to clean the hold, although we were bound for Callao to take in one of the dirtiest cargoes we could possibly have to handle.[6] This meant, of course, not only packing away every-thing movable but also the scraping of the internal lining. Now, this being a new ship and well varnished, it did seem ridiculous to me but as the sailors say "Obey orders if you break owners". If an order is refused they have the power to put a man in irons until he is willing to obey, or until he arrives at the first port, where he is then charged with insub-ordination and probably gets three months as well as losing his pay for the time he has put in. There is, however, no law against grumbling and that is declared to be an Englishman's right. Then the paint wanted scrubbing and this we were set to do by the First Mate, Andrew Smith. As I never like to be doing unnecessary work, I took the liberty of telling the Mate that I thought that they would probably find some more useful work for us to do before long. Little did I think at the time how true that would prove to be or the disastrous nature of it.

We travelled day by day in perhaps a 10 knot breeze running under lower topsails with the wind just abaft the Port beam. By the afternoon of the 10th of May, we had reduced sail to Foresail, Mainsail and Fore and Main Topsails. For the past two or three days we had been sailing by the compass and the log, that is to say under dead reckoning, as the officers had been unable to get a sight of the sun owing to hazy atmosphere. The log (the patent ones) had three blades at the end of a tube which contained wheels, dials etc., similar to gas meter dials and were read the same way. These give the distance traversed in knots and can be reasonably relied upon for accuracy.

*　*　*　*

British navigators first began to try to calculate the speed of their ships in the mid-16th century. A piece of weighted wood (the log) was tossed over the stern and the line attached to it played out. The length of line let out in a given number of seconds determined how far the ship had travelled. Later, to simplify calculations, intervals in the line were marked with knots which were spaced so that the number which ran out in the time of a sandglass equalled the number of nautical miles that the ship ran in an hour, hence the term 'knots.' Speeds measured in knots by means of a log were logged, or recorded, in a 'log' book. By the mid-18th century the technology had developed much as Holding described.

The *Mariners' Handbook* of 1921 notes that patent logs are "not always trustworthy".[7] The length of the tow line is an important factor and a log which runs accurately at one speed might not at a different speed. If the officers of the *Invercauld* were using incorrect data, perhaps it is little wonder that the Auckland Islands appeared when they expected no land nearby.

Navigational aids in the 19th century were still relatively primitive, charts were often inaccurate and the weather in this area is unusually foul. It was not an area to have to depend on dead reckoning. It is a known fact that the Aucklands did not appear on all the charts of the day, or if they did, were not necessarily accurately charted.

Henry Armstrong, in charge of the 1865 expedition to establish the castaway depots on the Auckland Islands, made the following statement in his official report to the Government of New Zealand:

> ...my attention has been drawn by Captain Thompson, harbour master, to a chart in the possession of Captain Boyd of the *Robert Henderson* to a chart published by James Imray, 102 Minories, London, bearing date 1851, on which the Auckland Islands are placed 35 miles to the south of their true position. This glaring error in the latitude will of itself account for the wrecks which have taken place in these islands, without the further aid of haze and current to which I have alluded in this report, supposing the masters of the vessels wrecked have been guided by Imray's chart. Captain Boyd is himself aware of the error, it having nearly cost him his ship on a voyage home from the Colonies. Immediate notice should be given to the shipping world of the existence of such an error on such a chart.[8]

It is quite likely that it was an Imray chart which was on board the *Invercauld*. Smith stated that the officers thought they were seeing the south end of the island, when in reality they were 35 miles (56 km) further north. This would be a perfectly reasonable supposition were they using the British chart.

The tragic irony is that if one plots the co-ordinates which Bristow charts on his map when he discovered the islands in 1806, they are only fractionally off from today's charts. A serious error must have crept in during the transcribing of information, with disastrous results. It is just possible that to protect valuable sealing grounds the position of the islands was deliberately misrepresented and the information from one of these charts was picked up by Imray.

The *General Grant* was wrecked almost two years to the day after the *Invercauld*, under almost identical circumstances. Eunson, in his book *Wreck of the General Grant*, discusses the lack of knowledge of the area.

If Captain Loughlin had been using the Australian Directory he would have had no information at all on the route. One might expect him to know the United States publication, but it may be that ordinary sailing ships would not carry such works as Maury's *Explanations and Sailing Directions*.[9] Nor do the new ones I have consulted go into detail about the route in relation to the Auckland Islands, so that even Maury might not be really helpful. Added to inadequate information on the Cape Horn route... was the difficulty of keeping an exact course because of the drift to leeward, currents etc. for which an average ship's captain might not be able to make due allowance. The fact that the currents around the Auckland Islands are particularly treacherous made it essential to give the islands a wide berth. This theory is heavily supported by experiences recorded in the logs of other ships' masters using the great-circle route. The southerly drift was such that several masters found themselves in much the same position as the *General Grant,* and while some escaped, others did not... Undoubtedly the same could be said with regard to Captain Dalgarno and the *Invercauld*. The captains of the *Derry Castle*, the *Anjou* and the *Dundonald* had all made allowances of between 30 and 40 miles (48 and 80 km) for the southerly set... Perhaps the greatest challenge to skippers sailing in this particular area was posed by the very powerful southern drift which could sweep ships miles off course.[10]

The wind had increased as the day went on, for according to Smith's account, by the time they sighted land they were sailing under the three closely reefed topsails only. The writer of an 1849 sailing manual states: "I believe I am right when I assert, that the main topsail of a sailing vessel should be kept set as long as possible, in blowing weather. More particularly when accompanied with a heavy sea. It is the resistance of the main topsail, placed on such a long lever as the mainmast head, which gives the ship that easy motion, which she so generally has when lying by the wind, under that sail."[11]

The reefed main topsail could not be braced up as sharply as the mainsail, because when close reefed, its yard would almost touch the lee topmast shrouds. As a result, it was impossible to bring the ship as close to the wind as could be done with the mainsail. The significance of this will be seen through Holding's and Smith's descriptions of the events leading up to the wreck.

At 4 p.m. on the 10th the captain had given orders to double the watch on the lookout, thus showing that he knew our proximity to land. We were then sailing on the Port tack with the wind just abaft the beam, steering S.E. by E. ¹/₂ E. The man on the lookout called out "Land on the Port Bow". From records I find that it was on the Mate going forward that the land was first reported and right ahead. He called out "Land Ahead". It was now twenty minutes to eight by the clock in the Mate's birth, as I saw it through the window. When he called out to luff,[12] it was pitch dark; so that it was impossible to observe the distance or the extent of it. Upon bringing the ship to the wind (to the port) and sharply bracing up the yards, we soon found that it was utterly impossible to get out that way, for in a few minutes they again called out in chorus, "LAND RIGHT AHEAD".[13] We now had orders to put her on the starboard tack.[14] And in a few minutes saw land almost right ahead so braced the yards up on the starboard tack. The order was to bring her around and brace sharp up on the starboard tack. We were thus leaving land astern.

After running that way for a little while we appeared to lose sight of the land. The officers appeared to think they were quite clear, so ordered the Ship around to her apparent previous course. We squared away the yards and that is undoubtedly where the fatal mistake occurred, as it was but a short time before we could distinctly see land right ahead.[15] Thus we fell into the land at about the same position as at our first sight of it, with a point on the port bow and the wind off the land. There then appeared to be only one chance of clearing the land – to get by it on the Port tack; tho' we did know there was plenty of open water in the direction of which we had come (to wit) the North-West.

I don't think we were more than three hundred yards from the shore. We soon found we were getting into a little open water, and with the wind sharp on the port bow, the yards braced sharp up, we passed a rock, perhaps two, on the starboard bow. These only showing a few feet above the surface, but quite enough to look ugly and dangerous.[16] We had now been pulling and hauling to the uttermost of our strength for probably two hours, but it was not over by a long way. We had not cleared the point when land was plain. Right ahead again and as far as the eye could see, all along on the starboard or lee side and even with bracing the yards up to the limit we had land on both sides. There then appeared only the one chance, if we were between two islands, we might get through; but owing to the extreme darkness, it was utterly impossible to have any hopes of doing so.

By this time everything had become confusion and all the officers appeared to have lost their senses. The captain was on the forecastle giving direction, the mate was on the main deck trying to obey, while the second mate was on the poop bullying the man at the wheel and the rest of us running from one place to another, hauling on tacks and sheets until our arms were almost useless. We had now braced sharp up on the port tack again and in a very short time the cry went up "Land Right Ahead" and we were so close to it that there was no mistaking it, and it was also plainly to be seen on the Port side, and it looked impossible to put her about in either direction. But land and water are both very deceptive at night. Had it been daylight there can be no question that we could easily have got out of that difficulty.

I say this from previous experience gained by a trip to the Murisius [sic]. We called at Port Lewis with passengers in Western Australia. The Captain of the ship "Marndine" was looking for the Pilot and going too close to the shore in a kind of bay had to go around quickly and in doing so I quite believe I could have jumped onto the rocks. He had plenty of room to come about but not enough wind.

While here, we had plenty of room but could not see it and we were now under the lee of the land and going very slowly and in this way we passed one or two rocks on the Starboard side and got into what looked like a bay.

The rocks just mentioned were only small ones and may not have been more than six or seven feet above the surface, but as we were thoroughly occupied in running from one rope to another we had not time to look round, even if it had been daylight. We were passing the land very closely on our port side, this we could not help seeing, it being so high and us so close to it. In a very short time it began to look as though we were getting into open water with land on both sides. The wind was now coming around a point and apparently right ahead of us as it will now appear we had missed the land on the port side but no one could say we were not running into a bay as the land loomed up high on the starboard side therefore nothing was to be done except to brace sharp up on the port tack. That is to say, tighten in the sheets, this we did until the sheets were as tight as boards and our arms were ready to break.

The officers had apparently entirely lost all hopes and also their heads, for the mate was crying, the Captain giving orders that could not be obeyed, and took to the boys while the Second Mate was bullying the man, Gogland, at the wheel to "Luff, Luff" while the sails were shivering all through. Is it then, any wonder that we were drifting nearly broadside on, instead of sailing? While if the ship had been given $1^1/_2$ or 2 points she would have been making headway. We were now about exhausted by the amount of hauling we had been required to do and had all sails like a board as the second mate still kept ordering the Helm Hard down.

The land was still plainly visible on the Starboard side as far as we all could see. The water was comparatively smooth as we were somewhat sheltered by the land on the port so it was possible to tell perfectly in which way the ship was going – which was probably about four points in front of the beam. She was therefore making dead for the land. We were now nearing the fatal rocks and by the water breaking into foam and the phosphorescent display, we could see our position more clearly. Of course we didn't know what was ahead of us at that time, but it was plain to us that we could not escape the land to the leaward which was now looming high and close, probably not more than half a mile distant.

* * * *

Before our trip, we had poured over maps and aerial photographs. We pushed tiny ship models, complete with moveable sails, this way and that along the coastline, trying to interpret Holding's sequence of the events of that terrible night. My husband Robin is a meteorologist as well as an avid sailor and he put all his powers of logic to work as we tried to second-guess the weather patterns of 130 years ago.

Our Scottish friend Stanley Rothney had put us in touch with Edward Irvine, an old friend now living in Australia. Mr Irvine is one of the few remaining skippers alive today who served in merchant sailing ships and he provided valuable insight about the actual mechanics of the timing involved in changing direction in a square rigger. This, of course, was a crucial factor in the wreck of the *Invercauld*. We, who are so used to motors, forget that it is not just a question of putting over the helm. He recalled that when he sailed on the *Rydberg*: "...a big Scottish-built four-masted barque, then under the Swedish flag. We invariably tacked ship with all running gear prepared beforehand. The total operation taking perhaps 10 to 15 minutes depending on the wind and sea conditions. Neither the *Killoran* nor the *Pamir* were ever tacked in my time but were always worn (gybed) around. Perhaps a safer move in a tight corner, sea room permitting, but it took about twice as long as tacking."

Of course gybing in a square rigger is totally different from performing the same exercise in a sloop. In the latter, when the boat comes up into wind the sail luffs and when the boat comes round onto the other tack, the wind fills the sail almost immediately with little loss of headway. It is much more complicated on a square rigger. Here, when the wind gets onto the back of the sail, the boat is pushed backwards, and the whole angle of the sail must be changed.

In proof of my statement of the demoralized state of the officers, I will now endeavour to prove by actual experience that this is correct. The captain gave the order to "Cast the deep sea lead" and as it happened I was there. I don't know how I found the Lead but I did find it in short order. I knew that it was useless to try to cast it on the Lee side as is usual so I cast it over the Port side, where instead of the line trailing Astern, it trailed nearly direct on the beam. I don't know if it ever touched the bottom, for we were then so close to the rocks that nothing else mattered. Before I had time to communicate the result the captain in his delirium called out "Drop the Anchor" although, as I have before stated, it had been Catheaded five or six days and cable put below and no preparation had been made for this.

This should show all interested parties the result of lost nerves when most required; in all deference to the men concerned I feel that I ought not to pass these things over without making some statement, as it did certainly contribute greatly towards causing the wreck and to our Future suffering.

He then gave the order to cut away the grips of the boat. We had on board three boats, The Long Boat, the Captain's Gig, and the Panace, I never knew what became of the two last named. The Long Boat was, as is the usual custom, turned upside down on the house on deck and lashed there. I was there again and always had a good sharp knife, which I always pride myself in carrying and may I here state that it was the means of saving a young man's life some twelve years later. I had cut one or two of the lashings and when no one had come to help, and on looking for them there was no one to see, it only took one glance to see the reason – I had been so absorbed with the object of clearing the boat that I had not noticed our position. We had now drifted into a kind of chasm, or bight, with towering rocks fore and aft. It was only likely to be a few moments 'ere we struck the rocks broadside. Seeing how close we were to the rocks then left it and don't know that I ever saw it again.

The wind was blowing at an angle of about 45° and with all sail standing we drifted in with the seas breaking all around us. No one ever knew how long it was since we sighted the land, but I should imagine from two hours to two and a half, though it might have been longer. That would bring it to about 10 p.m. From that time I had not been in front of the Main mast except to haul on the Fore sheet, with the consequence that the Poop seemed the safest place to me. We still drifted in slowly as we appeared to be somewhat sheltered from the wind by the land on Port beam but time was so short before we struck broadside on that the only thing we thought of was, would we ever be able to land in such a place, for it looked the same on three sides and the only outlet was by the way we had come in.

The Foresail, Mainsail, and lower fore and Main topsails were still set and stretched like a board, but the Gibb [sic], which had been spread, had gone together with the Gibboom [sic] – the latter had been carried away when she struck the rocks. I never knew when or how the boats went.

It was quite plain to every one that there was now no escape for us. Discipline, which had been very lax for the last half hour, had now failed entirely. There were no more orders given nor were they necessary. It was now every man for himself with the most dismal outlook anyone could possibly imagine.[17]

*   *   *   *

Captain Dalgarno's account, not surprisingly, is somewhat different:

The fog, which by nightfall grew denser and denser in the vicinity of land, had prevented us seeing them [the islands] sooner. Suddenly the N.W. breeze was replaced by a dead calm, and during this we were at the mercy of the strong currents which render the approach to the Aucklands very dangerous. Imperceptibly they carried us closer inshore.

The rapid fall of the barometer disquieted me greatly. Soon after sunset the sky was overcast with thick black clouds, which indicated bad weather. Towards midnight a violent gale broke out all at once from the south-west and placed us in a very critical position... We crowded on the ship all the sail she could carry, but in spite of our efforts I soon saw that she was destined to perish on the rocks. At 2 o'clock in the morning a frightful shock sent both our masts by the board. The fatal moment had arrived.

The Invercauld had struck upon a reef near a lofty cliff. Close at hand a little cove, where the rocks were less elevated, attracted all our attention. It was useless to think of saving the vessel, which was soon dashed to fragments by the breakers. I succeeded in swimming to a little cove, where I clung to the rocks with all my remaining strength. Some of my crew, who had got there before, helped me to reach the shore.[18]

[1]   Noble, J., *Port Philip Panorama: A Maritime History*: p.55.
[2]   ibid: p.58.
[3]   Austin, K.A., *Port Philip Bay Sketchbook*: p.45.
[4]   ibid: p.48.
[5]   To secure.
[6]   Guano was picked up on the Chincha Islands, two days' sail south of Callao, the port for Lima, Peru.
[7]   *Mariners Handbook*, 1921: p.79.
[8]   New Zealand Government Gazette, Southland, April 11, 1868. Vol. 6, No. 9. p.56.
[9]   Maury criticized the *Australian Directory* for the fact that it did not include information on the fastest route – via Cape Horn. Maury's directions however did not show the Auckland Island even though he gave directions for the Cape Horn route.
[10]   Eunson, K. *Wreck of the General Grant*: p.104–109.
[11]   Hartland, J., *Seamanship in the Age of Sail*: p.216.
[12]   Bear towards the wind.
[13]   Disappointment Island.
[14]   To head N.W.
[15]   Smith: "We had not run long when to our wonder and astonishment, land was again reported right ahead."
[16]   Invercauld Rocks?
[17]   See Appendix Three.
[18]   Raynal, F.E., *Wrecked on a Reef: or, Twenty Months among the Auckland Isles,* p.328-329.

6

New Zealand

Thomas Musgrave.

Stone and sand and sea and sky
Rest my heart and please my eye.
I will go and not ask why
Stone and sand and sea and sky.

Rose Vaughn

AUSTRALIA slipped beneath the horizon as our plane turned eastward towards "the land of the long white cloud". Joined by our daughter Brenda, we followed spring southward for the next six weeks, meandering though New Zealand's sheep-studded countryside towards our rendezvous with the *Evohe*.

Manapouri, a half-day's drive from Invercargill, is a cluster of houses nestled on the shores of a mountain-fringed lake. Here, the Shaws have a bed and breakfast and when we at last drove up to Home Street we were welcomed with open arms by Ruth. Lance was at sea, so it was several days before we met him, but after all our letters and faxes, we felt we were coming to visit old friends.

The week that followed was one of peaceful relaxation. Robin and Brenda headed off to the Milford Track while I finished preparations for the Aucklands journey. Each evening I slipped away with my fly rod, to stand by the side of a stream and wait for the evening rise.

Our project had elicited little response back home, but the New Zealand media was certainly interested in these mad Canadians. After an article in the Auckland paper had been picked up in Christchurch, a descendant of Alexander Barnes, the steward of the *Invercauld*, contacted us. On our way back through Christchurch we spent an enjoyable evening with him and his mother. Alexander Barnes was no longer just a name on a crew manifest. One more reason for a permanent memorial other than the lonely grave of James Mahoney in the tiny Hardwicke settlement graveyard.

Our first sight of the *Evohe* was from the deck of a Fiordland Travel tour boat. White and sleek, she lay moored in a tiny cove beneath the towering cliffs of Doubtful Sound. Through the eyes of a landlubber she looked incredibly small to take on the fearsome seas of the Southern Ocean. The words of Musgrave, writing about the Aucklands, came back to me: "I have been around both capes (i.e. Cape Horn and Cape of Good Hope) and have crossed the Western

Ocean many times, but never have I experienced, or read, or heard of anything in the shape of storms to equal those of this place."[1] What in heaven's name did we think we, a camp director and a university professor, with two grown children, were doing, risking our lives on a sentimental historical quest like this? But the *Evohe* was a solid sea-going vessel and Lance, who had been skipper of a DoC research vessel for 12 years, had an excellent reputation as a careful and thorough seaman. We would be in good hands.

It was finally on to Invercargill where, thoroughly jet-lagged, our son Bruce and my brother Jim were waiting for us. Now that we were all together, we were even more anxious for the adventure to begin. It was incredible to think that 10 years ago I didn't even know I had a brother, and now we were setting out to follow our great-grandfather to the sub-Antarctic.

The first stop was the Southland Conservancy offices where after four years of correspondence we finally met Pete McClelland, our "Man from DoC". The inevitable paperwork out of the way, we headed to the Southland Museum to see the sub-Antarctic exhibit. Sitting in the darkened theater, I felt like a child with my nose pressed to the window of a fantastic department store Christmas display. The marvels of the Auckland Islands – images of sea-lions, soaring albatross, strutting penguins, waves crashing on impossible cliffs, ancient drawings of shipwrecks flung across the screen. The mantra, "we're going, we're going," pounded in my brain. My dream was becoming a reality.

We awoke on the morning of departure to a howling gale from SW with heavy rain. Lance had told us that he didn't want to see us at quayside in Bluff before nine that night as he didn't need the extra headache of passengers milling about during his final preparations for sea. We spent the morning dealing with inevitable last-minute details and with a brightening afternoon sky above us we piled into our overloaded station wagon and headed down the highway to Bluff. This windswept seaport is a bleak cluster of houses clinging to a rough hillside – certainly not a tourist town. We drove down the main street looking for a restaurant but very soon ran out of road. We stood at the rocky southern tip of the South Island and stared across Foveaux Strait in the direction of Stewart Island. The inscription on black rock at the foot of a bright yellow signpost showing distances to Singapore, Tokyo and London reminded us:

MIGHTIER
THAN THE THUNDERS OF MANY WATERS,
MIGHTIER
THAN THE WAVES
OF THE SEA,
THE LORD ON HIGH
IS MIGHTY –
Psalm 93, v.3/4
God is always greater than all of our troubles.

The Stewart Island ferry swung out of the harbour, looking minute as it disappeared in the troughs of the waves. Were we really going to set out in this? Each of us concealed our thoughts behind family banter.

The clouds closed in again as we retraced our steps back to town, praying that we had missed a restaurant somewhere along those bleak streets. A tired blue awning swaying in the wind was the clue which led us into a lovely old hotel. Closeted in the cozy bar, the fireplace throwing out welcome warmth, we toasted our expedition – and ourselves – and our boat – and the gods of the sea. We dined in style, making the most of the extensive menu and wine list.

By the time we emerged the wind had increased and an angry yellow-black sunset cast an ominous glow over the harbour to the hills beyond. Once through the port gates we drove past huge stacks of containers, piles of logs destined for Japan, fishing boats, container ships – everything but something which looked like the *Evohe*. Then as the next squall hit with a vengeance, we spotted the tips of two swaying masts on the seaward side of the furthest pier.

Sitting in our violently rocking station wagon, buffeted by wind and rain, our conversation was heavy with gallows humour. When the worst of the squall passed, we tentatively opened the doors and, hanging on to keep them from being snapped off in the gale, five valiant sailors slipped out into the rain. Bruce, a little braver than the rest of us, went to the edge of the quay to look down. Another sudden gust of at least 80 mph blasted us. I watched my son blown sideways until he was teetering on the edge, balanced on one leg, with a 10 foot (3m) drop to the deck below.

There she lay – windswept and wet with nothing so modern or advanced as a ladder, just a large bald truck tire chained to a bollard. Bucket brigade style, our gear was passed down to the crew. Getting ourselves on board was much more interesting: a long stretch down to the tire, then with one hand clutching the slippery chain, the other reaching out to a helping hand on the boat, the step of faith onto the wire railing and down to the 'firm' deck. Single file we squeezed between wire and the wheelhouse to the afterdeck housing. My brother stopped, pointed an accusing finger in my direction and shouted over the howling gale, "You're nuts!"

Amidst the confusion of finding cabins and stowing gear we made our first acquaintance with the crew – Eric Ashton, a local New Zealand lad, mate and cook; Mickey Squires, a Stewart Islander from Half Moon Bay was sailing as our second skipper; Les Hutchins of Fiordland Travel, engineer. Our 'deckies' were the 'gentle giant' Prebin Wolfe from Denmark (or 'Mr P', as Brenda immediately christened him), and Peter Harford-Cross from Yorkshire, England. We were missing the photographer Paddy Ryan and Pete McClelland, who had both been warned that there was no way we would be getting away on the early tide.

Robin and I stowed our gear in the second cabin on the port side. Pulling open the narrow cupboard, two lifejackets – bigger than any I had ever seen, glowered from the semi-darkness. Across the companionway, Brenda spread out in the other double cabin. Two more cabins, two heads (washrooms, complete with tiny tubs) and the galley completed the forward area. A roomy lounge area filled in the centre portion, then five stairs up to the wheelhouse. The captain and crew quarters were aft. This was the *Evohe*. Wonderful!

We crawled into bed just after midnight with the sound of the wind sweeping round like a mad creature, the shrouds clicking and the rain slamming onto the portholes. Even though we were still firmly tethered to the Bluff quay, there was a definite movement of the boat and my diary entry for that day finished with this sentence: *I refuse to be seasick while the boat is still tied up!*

Morning brought no change in the weather – things did not look promising for getting away that day. We could only keep our fingers crossed that we would not be delayed too long, for we had budgeted for 10 days, maximum 12, and the *Evohe* had other commitments, as did our bank account. It would be devastating to have to scratch the voyage after all this preparation and expectation. The forecast called for the wind to drop and seas at a maximum of five meters (16 feet) – which sounded quite high enough. We used the time to go into Bluff and Invercargill for some last-minute important items: white spirits (for the stove), more wine (hope springs eternal) and more seasickness tablets. Pete came aboard, followed shortly by Paddy. To know him was to love him and his collection of the worst jokes anyone ever told.

By early afternoon the skies began to clear and despite the high wind we motored over to the fuel station and 'tanked up'. No friendly attendant, just a hose snaking down from a gull-crowded dock. Lance decided it was worth trying to make the run across Foveaux Strait and down the coast of Stewart Island. If the system hadn't passed by then we could hole up in Pegasus Harbour with seven hours sailing under our keel. We finally headed off on the afternoon tide at 4:30, only 12 hours past our ETD.

We chugged into the centre of harbour while the Bluff tugs performed their 'mating dance' as they shepherded a large freighter into its berth. Lance pointed the bow seaward; we slipped past the tiny fishermen's houses clinging to the hill, past the harbour light and civilization faded into the sea mist.

Sails taut beneath a leaden sky, the *Evohe* plunged through the grey-green waves of "that waste of waters that is the Foveaux Strait."[2] Pitching and rolling, we ran down the rugged, unpopulated coast of Stewart Island. I was too excited to think about being seasick. My husband, an avid small boat sailor, sat glued to the small seat beside the radio. Not to clutter 'the office', the rest of us came and went, alternating our time between balancing in the wheelhouse and wedging ourselves on the benches below. As I watched the succession of grey waves sweeping towards us, passing under the ship and continuing on – to die somewhere in the Pacific – I pondered the long, long road that had brought me to this point in my search for Robert Holding, seaman.

This was, in essence, the second part of many years of searching; the first search was for my birth family. The reunion was joyous and fulfilling, and resulted in a blended family. My adoptive family had given me a great sense of the importance of family history; now I had a whole new family to relate to and discover. Part of this discovery was the manuscript. But it was so much more than just a family search, far more than taking the 'adoptee syndrome' of rediscovery to an extreme. Life is to be lived, and I was tired of being a workaholic, tired of the routine of life. Here was the chance of a real life adventure, here was the opportunity to go to the ends of the earth, with a purpose.

After our trip was over, I heard a song which expressed the feelings I could not articulate:

Stone and sand and sea and sky
Rest my heart and please my eye.
I will go and not ask why
Stone and sand and sea and sky.

Soon the wind is holding me
Clears my mind so easily
Open, open to the song
Wind and sea have played so long.

I am strengthened by the sea
Something broken mends in me
Hold me till the day I die,
Stone and sand and sea and sky.

Stone and sand and sea and sky
I am free to laugh and cry
I feel the spirit lift me high
Stone and sand and sea and sky.[3]

As dusk was falling, our skipper turned and said: "We're going for it, no point in stopping." We were on our way south.

[1]   Musgrave, T., *Castaway on the Auckland Isles*: p.89.
[2]   Carrick R., *New Zealand's Lone Lands*: p.4.
[3]   Vaughn, R., Stone and Sand, from *Fire in The Snow*.

7

Wrecked

heavy confused sea

It was quite plain to every one that there was now no escape for us. Discipline, which had been very lax for the last half hour, had now failed entirely. There were no more orders given nor were they necessary. It was now every man for himself with the most dismal outlook. Our feelings can be better imagined than described.

With black rocks on three sides and a raging sea on the other, there was no chance of launching a boat. The only thing that I thought of was to get a footing on the rocks. As I ran aft with this idea in view, I well remember one of the men, a Londoner by the name of Tom Page who had been sick a few days, calling out to me, – "Bob, for God's sake, don't leave me". Well, what could I do except tell him to follow me and that I would do the best I could for him. Poor fellow. It proved to be very little assistance I could render to him, as will shortly appear.

He did not appear to have followed me, so getting on the poop and looking over the Stern, which was resting on some rocks just under the starboard quarter, I thought I could see my way to escape from the wreck which was fast breaking up. I got a rope and slid down over. To my great surprise I had landed on a rock which was separated from the main rocks by perhaps ten or fifteen feet of water. To make things worse I was amongst a tangle of seaweed resembling greasy sea serpents. They had great oval leaves on branches as thick as my leg and I don't think I am wrong when I say they were up to five fathoms in length. I don't think I was more than 20 feet from the rocks on which I wanted to get a footing, but would have to pass through about fifteen feet of water and this tangle. Say it took but a short time to convince me that it was impossible to get through that way with safety. I had been able, while on that rock, to notice that there were a lot of loose rocks and boulders which had fallen from above.

I found I was unable to get back up by myself. No one had seen me go over but I heard some of the men talking on the poop and I called out for assistance. Fortunately they had been able to hear me and I was soon hauled up by two or three of them. There were now six of us (Thos. Turner, Thos. Page, Jas. Tait, Wm. Gogland, and one whose name I do not remember) on that end of the wreck, as "wreck" she was surely then. The bowsprit had been carried away while passing in, owing to the width of the passage being so narrow and the length of the ship so near equal. She had in the first place canted in to Starboard, but now the ballast had gone through the bottom, the waves were lifting her up onto the rocks and the decks were broken up. We could not now see more than half the length of the ship, what we could see was awash and settling down very fast. While we were trying to take in the situation, she took a list to port and the masts were carried away. There was no possibility of getting to the fore part. It was with difficulty that we could stand and we took ourselves to the Taffrail and the rigging for support. It was, however, only for a few moments as the Mizzenmast soon followed the others, bringing the rigging down across the deck. Tom Turner had got his arm jammed between the rail and the deadeye and the lanyard. He called out to me, "For God's sake Bob, cut the Lanyard, my arm is broken." I always carried a sharp knife and he was soon released and we were glad to find his arm was not broken.

The water was now coming in lumps. There were now only about fifteen or twenty yards between us and the rocks, the space between being covered with floating wreckage. Everything now was nearly level with the water except the starboard rail, the rigging – to the maintop, and the part of the poop where we were standing. The whole bottom appeared to have fallen out of the ship. The seas began to come in heavily so I called out to Tom Page and the others to follow me. (It may here be explained that the Maintop is a wooden structure for the support of the Topmast Rigging, although not in reality the top, but is to support the topmast or the second from the deck. The Fettic Shrouds pass through to Deadeye for the Topmast Rigging Attachment.)

I went along the rail to the Maintop and just got my breast against it and my arms around the rigging when a heavy sea struck us. Being anxious for the others as well as myself, I looked round to see if they had held on, but to my horror I could only see two instead of five. There was little time to lament as there was another topper close to me. So, bracing myself, I tried to hold on as long as possible – and was none too soon as it struck with a terrific force. As was natural, I then took another look for my mates and alas, not a soul was to be seen. I was now alone as far as I knew, but again there was no time to lament as there was another coming fast. This was to be expected as it is a well recognized fact that there are always three large seas which follow each other in quick succession.

I had only just time to take a good hold when it struck with such force that I never knew how I let go my hold. I thought the others were heavy, but this one took all the breath out of me with tons upon tons of weight. My only thought was to get my head above the surface to get a breath of air which I soon accomplished by hard struggling. I found myself in the water amongst all sorts of timber and broken planks and almost before I was aware of it, I felt myself on the rocks. No sooner did I feel my feet there than I began to wonder if any of the others had been saved.

It was not much trouble to get onto the land for the rocks around the wreck were of all sizes. Once getting up clear of the water, except for the spray (which I had no need to mind, being thoroughly wet), it may be truly assumed that I was anxious to know if I was alone in this terrible plight. I called out with great hoots in the colonial style, and was delighted when I heard a reply. One by one they came round to me and all were as pleased as myself. After a reasonable time we took a roll call and found we had mustered altogether nineteen souls: The Captain, George Dalgarno; Andrew Smith, the first Mate; Jas. Mahoney, the second mate; the Carpenter (Scotch), the Boatswain (nationality and English name unknown), the cook (Spanish), the Steward (Scotch), Ten Able Seamen – the names that I remember were: Thos. Turner, Wm. Hipwell, Jim Tait, Jim Harvey, Dutch Peter, a fellow we called Fritz, myself and the boys Liddle and Lansfield (Scotch). Thus, unless more turned up we had lost six, with poor Tom Page amongst them. Of course we were in hopes of others turning up when daylight came.

Many who had escaped by the front of the ship had reached the rocks quite dry, that having been the most sheltered position which had been overlooked by me. We crawled amongst the rocks to find shelter and huddled together but sleep was completely out of the question. Being thoroughly wet I was glad of a little protection. The wind appeared to have increased and the spray was continually breaking over us, making it one of the most miserable nights that any one ever suffered. There was not a watch amongst us which was unfortunate, for in such a situation the knowledge of time can be solacing. The only thing to do now was to wish for morning so that we could realize our position more clearly.

We could now do nothing but exchange our experiences and congratulate each other on our escape so far. Some reported having seen one of the crew, who had been struck with something on the head, hanging through a break in the forecastle. As for the five who were washed from the poop, only Tom Turner was to be found.

When daylight broke on the Eleventh Day of May it was possible to look round and try to find out our position and it did not take long to convince us as to the deplorable position in which we were placed. This is rather difficult to describe, however I will try to do so. The seas were still breaking heavily and we were kept continually wet by the spray flying high over us. The scene was that of a forlorn hope for the cliffs behind us were enough to shake the courage of the stoutest heart.

Where we had landed and spent the last part of the night (or early morning), was on a pile of loose rocks which had undoubtedly fallen from the face of the cliff, probably during some ancient earthquake. At our back we had a heartbreaking sight. The rocks were of various sizes from small ones to great boulders of three or four tons. They were piled up near the face of the cliffs to a height of probably 200 feet above the surface of the water and sloped down at an angle of something like sixty degrees. The cliff rose nearly perpendicular with broken patches here and there. This was to us a terrible sight. I don't believe there was one who ever thought we would be able to escape from there.

Looking out to seaward, we had on our left a high precipitous bluff which appeared to be double the length of the ship. This we thought might be from five to six hundred feet high running in almost a straight line and rising as it passed into, and connected with the main land. At the extreme outer edge of this formation there was a rock about six feet high and nearly four feet square standing on one end, which I shall have cause to mention later on.

We had on our right, a ridge of rocks of broken formation running from the water's edge at the entrance, continuing up to probably one to two hundred feet. At the back of this ridge was a chasm or gully – narrow at the top and widening out to about 40 or 50 feet at its termination near the opening. This we thought would be three or more hundred feet high. The face of the rocks looked the same as the ones on the left, except that half way from the head of the Chasm there was a sloping part about a third of the way out (about 15 to 20 feet wide at an acute angle), where there was some herbage up to about two feet high and a few small bushes.

There were also a few small patches on the luff which I took it to be celery growing wild. Here there was a rock about two hundred feet up that I shall have something to say about later.

Having explained the appearance of this inhospitable place to the best of my ability, I may now endeavour to give an idea of our feelings and how we passed the day. The reader may from the above description try to formulate some idea of the state of our minds while placed in such a position, for it is impossible for me then or now.

I have stated that we had not a watch amongst us, therefore the time did not trouble us much.

8

Lat 51° 8' S.; Long. 186'

And all I ask is a tall ship,
And a star to steer her by.

John Masefield

MY experience with boats had extended no further than canoeing and sailing on lakes and various ferry crossings. My excitement and anticipation of the trip to the Aucklands had been tinged with a fear – I had been afraid of being afraid and afraid of being seasick. However, despite 9 to 12 foot (3–4m) waves, a 30° roll and a rather interesting pitching motion, there was simply too much to see (even in the middle of the ocean) to allow myself to be either. I was simply overwhelmed by the sheer joy of being at sea in a small boat. The emotional experience of being under sail, far from the sea lanes, with nothing but sea and sky and soaring birds drove any other thought from my mind.

I had taught geography classes for years that the Pacific Ocean covered an area greater than all the continents combined,[1] but like my students (who dutifully wrote down the facts to be memorized), I had never truly absorbed the reality of the vastness or power of the sea. I had stood atop towering cliffs, mesmerized by the surge – but when your feet are firmly on dry land the perspective is very different from that viewed from a heaving deck.

Modern life does not prepare us for the experience of real isolation. I am one of the fortunate who have lain beneath a star-studded sky in the northern wilderness, the canoe by my head the only link with civilization – but nothing, nothing, could compare with this. East to the Cape of Good Hope, west to Cape Horn – between – emptiness, except for the waves on which we rode. An endless expanse of sea on all sides. Great swells lifted us towards the soaring albatross, then flung us down. The only sounds were the wind in the rigging and the solid 'chuck, thunk' as the ship took the waves.

The grey light from a grey sky was reflected onto the sea, creating a luminous effect which I could almost feel. In the same sense that the eye becomes sensitized to the infinite shadings of blacks and greens and whites of a Canadian winter when we become so starved for colour, so we became conscious that the sea possessed an infinite palate of greys and blues. The tones of white in the spume blowing from the wavetops were always shifting, each wave different from the next, or the last, as it flowed under our ship. I stood for hours watching the sea.

Shortly after we had reached the open sea, the albatrosses became our constant companions. No words can describe the grace and power of these awesome birds as they ride the air currents. With wingspans up to 11 feet (3.4m), they would soar upwards until they were only a tiny speck against the blue and then gracefully, slowly, gently, they would lean into the wind and swing down and down until they were skimming the waves beside us, another tilt and they were gone again. Despite the common belief that they never land on the sea we saw one touch down and then rise again into the air just as the wave reached its crest. It simply spread its vast wings, caught the wind and was airborne.

The little brown and white cape pigeons were also very much in evidence. Jim described them "like fighter aircraft swooping in and around and up and over the sails. They swing back around and make another run at us."

Living as we do in a world of speed, life aboard a sailing vessel offers a different reality. We think of the trip from Toronto to Montreal to be a considerable journey, yet at 120 km per hour, it takes not quite six hours. We hop on a plane and in the same amount of time travel from Montreal to London, England. Now we found ourselves commenting on what excellent time we were making as we 'rocketed' along under sail at six to nine knots.

I enjoyed every moment aboard. At one point Lance stuck his head down the hatchway and observed: "You are doing nothing for the image of the strong New Zealand seaman! My engineer is crook, your son is crook, the DoC man is crook, and there you sit, having eaten an enormous breakfast, reading and typing – which I told you would be the worst things you could possibly do – with a grin on your face and waiting for lunch." I felt rather proud of myself. Never once in the whole voyage, except for the first quarter hour back on land, did I feel queasy. I guess the old boy's blood, though diluted by the generations, still flowed true. Fortunately I was never put to the test by the fury of a storm like the one that had doomed the *Invercauld*.

No one went onto the afterdeck without permission, but the waves and cold wind were such that we did not feel penned in by the heavy steel door. We were always welcome in the wheelhouse and the only understanding was, "if I have to push you out of the way, I will apologize later". We were very appreciative of Lance's patience and understanding in having his 'office' littered with sightseers. Our experience on the *Evohe* was quite unlike a well publicized Auckland Islands voyage on another vessel where the journalist was handed a

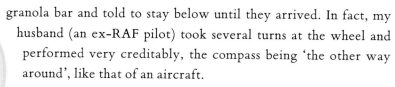

granola bar and told to stay below until they arrived. In fact, my husband (an ex-RAF pilot) took several turns at the wheel and performed very creditably, the compass being 'the other way around', like that of an aircraft.

By evening the seas were still running heavily. The expression 'getting your sea-legs' is quite apt, for learning to move comfortably from place to place was an intriguing challenge. The companionways are narrow enough that with judicious use of elbows and hips it is possible to stay on a relatively even keel. To get from a sitting to a standing position, not to be thrown back into the seat just vacated, took a modicum of judicious timing. Changing position in the saloon was an acquired skill: eye fixed upon the destination – body aimed in the general direction you wished to go – the moment of launch chosen precisely at the instant the ship began the roll to the side you wished to go – lead with the foot opposite to the pitch – then project yourself forward with enough force and speed to arrive at the desired location before the next roll took you back from whence you came. For safety, it was a good idea to have arms outstretched, like a calculating drunk, and trust that you had judged the roll correctly. Naturally, you did not change position unless absolutely necessary.

How Eric ever managed to come up with such appetizing meals under those conditions, I will never know. Chow was called and most of us arrived, even those who had been catching up on sleep at odd hours. With the exception of the man at the wheel, we all wedged into the saloon; a close neighbour helped to keep you on your seat. The end seats tended to be the most popular, but it depended whether you preferred to eat going back and forth, or up and down. Eric would appear at the foot of the steps from the galley carrying two, sometimes three, plates at a time – miraculously managing always to deliver these right side up.

Although we had a sextant on board, it was used more as a conversation piece and to instruct our young seamen than for navigational purposes. However, we felt that Mickey was somewhat more relaxed after he had 'shot the sun'. I was quietly delighted that a 'tell-tale' ribbon, familiar to all small boat sailors, flew from the forestay despite the fact that the bulkhead of the wheelhouse was filled with the most modern navigational and radio equipment.

The GPS (Global Positioning System) is probably the greatest recent advance in navigation. Taking bearings from a network of satellites, you can plot your position to within a few meters. Usually, between three and seven satellites are being used at any one time for a fix. Once the latitude and longitude of the destination are punched into memory, the ship's position is constantly updated. Any drift or leeway is automatically recorded and if required, a new heading is displayed on the monitor. Whereas the early ships had to depend on the patent logs, the GPS displays the boat's speed as 'over the ground speed', and thus any necessity to try to calculate the differences between speed over the ground, and speed over water is removed. The speed of the boat is constantly calculated by the changing position of the vessel in relation to the tracking satellites. Should the dreaded cry "man overboard" be heard, the helmsman would hit the emergency button and the GPS would then record, and hold, that exact spot. In the time that it took to put a ship about, a person would be lost in the towering waves but with the GPS system it is possible to return to the exact location to begin the search, greatly increasing the chance of survival.

What a change from the days when seamen had to depend on tracking the heavenly bodies. As the men of the *Invercauld* discovered, to their peril, days would go by when sights could not be taken, and positions were only guessed at by dead reckoning.[2]

* * * *

Diary: Thursday, 2 Dec., 11:30pm: "Here behind the stormsheet, in bed fully dressed. Have to get up when Aucklands appear on radar – they promised to call me. Fair winds today – picked up an easterly. Went for a bit without engines and dropped only one knot. Albatross, petrels around the boat all day. A day to relax (despite the necessity of always being wedged in – somewhere). Read and reread the manuscript, rechecked our plans – so much will depend on the weather. Expect to make landfall 3:00am."

All the bunks were equipped with a heavy canvas sheet fixed to the board underneath the mattress and attached to the deckhead by massive ropes and clips. The slight feeling of claustrophobia was offset by the security of knowing that you wouldn't find yourself thrown against the opposite bulkhead or onto the floor. Only once did I wake up to find myself lying on the stormsheet.

The call came at 1:30am. Instantly awake, I swung out from behind the stormsheet, and aware of much less movement of the ship, clambered up to the wheelhouse. Lance's tired face was lit by the soft amber glow from the radar; apart from the one tiny light over the chart table, everything was dark. Mesmerized by the circling finger of the radar, I watched the green blob that was the Auckland Islands, grow almost imperceptibly. The clouds were breaking and a thin moon cast eerie light over the swells. Thirty-three hours after leaving

Bluff, Lance brought the *Evohe* up into wind and the guttural rumble of the engine cut the silence.

The powerful masthead light was snapped on, and as we cautiously motored towards the dark, linear smudge the men lowered the sails. The single green blob in the centre of the radar screen broke up into identifiable islands – first Enderby which soon, amoeba-like, split as Rose Island became an entity. How different our feelings of eager expectation from the terror which these same islands must have engendered in those who had unexpectedly seen land looming before their ships.

As I watched the sweeping finger of the radar I was poignantly aware that 130 years ago my great-grandfather had stared out over this same stretch of sea, waiting for rescue. Now, here I was motoring slowly towards the safe harbour of Erebus Cove. Despite the passage of time, I felt the bonds of kinship that had brought me to these islands, the place where I, like Grampa Holding before me, would experience the greatest adventure of my life.

I tore myself away from the radar and looming shape of the hills, and crawled back to bed. At some point I was roused from my dreams by the 'chunk' of the anchors going over the side and the sudden silence as the engines were shut down. A few voices, then the ship was still.

---

[1]   The Pacific Ocean covers an area of 64,186,300 square miles; world land area is 57,900,000 square miles.
[2]   Dead reckoning: "A calculation of a ship's position by astronomical observations and reference to logged information." (Webster)

9

Shipwreck

Mollyhauks

*"Scientists used to assignments in the remote Antarctic and Arctic have called
New Zealand's sub-Antarctic islands the most inaccessible in the world."*

Conon Fraser

RUSTY-brown, the great hills swept upwards a thousand feet into a sky still flecked by dawn. The volcanic plug on the top of Mt Eden a beckoning finger. I stood alone on deck, filled with the wonder of my actually being in the Aucklands. The *Evohe* rode gently at anchor on a black, mirror-like sea, protected by the giant breakwaters of Enderby, Rose, and Friday Islands to the starboard. Shoe and Ocean Islands lay port side off the stern. Albatrosses, mollymawks and nellies soared above, while a curious, tubby brown and white cape pigeon meandered around the hull. It was sheer magic.

I hadn't realized it at the time, but my brother Jim had been up long before and recorded his special moment.

"5:15: The air temperature is, I would guess, about 45°[F]. I see a very heavy forested shoreline that appears to peter out towards the more mountainous area in the back. Very rugged terrain – very, very rugged. This bay is almost horseshoe shape. It is *so* peaceful; there is not a sound – no cars, no trains, no nothing, just the wind and the creaking of the masts."

We were only too glad to be on the move to see if there was anything to be saved from the wreck. Food was the main object so we spent the day in looking for anything that might have got washed up from the wreck. Upon searching round we picked up about two pounds of pork and about the same weight of biscuit which was a godsend to us. We tried to find others who might have escaped, but to no avail. The only part of the ship still visible was the stern which was landed on the rock that I had been on the night before. The front part was in the water and the cabin was entirely submerged. The deck, pointing upward at a high angle, had been stripped bare. We had hoped to find some sails and ropes but there was no sign of either. This was attributable to the fact that the ship had iron masts and wire rigging and the weight had carried everything into deep water. Wreckage was beginning to pile up and little was to be seen. Near the rock on which the stern rested, was a cave perhaps six feet wide and four feet or so high at high tide with a semi-circular roof.

Some of the wreckage having got washed up contained a few planks so we tried to make some sort of shelter. Using a hole between the loose rocks we placed the planks on one end around three sides and with others on the top of them we managed to make a place perhaps eight feet long by five feet wide and perhaps five feet high, with plenty of ventilation all round. This formed the only shelter we ever had there. It did help to break the wind somewhat tho' it eased our plight but little. Throughout this day there was little conversation between us for all were too dispirited to talk much at that time.

Although the wind was still blowing hard it did not affect us to the extent that we might have expected, owing I think, to the height of the rocks at our back. This being winter time coming on in that locality, the days were very short. We passed through the 11th day of May, one of the most dismal days that anyone ever suffered. There was however, one bright side to the matter. The Steward, who had fortunately escaped, had been lighting the lamps the night before and had luckily slipped a new box of wax matches in his pocket where they were safe, though a little damp. These were a godsend to us. The cook likewise had a round box nearly full with wooden matches in his pocket. He tried to dry his and to our disgust burned the lot. The mate then started to dry the wax ones and they took fire also but being near I swiped them out of his hands. It was but a moment 'ere the lid was closed, but not before too many of them were destroyed but I had saved the bulk of them. I may here state that neither the mate nor any other ever handled the matches again. Furthermore I have today, (January the 6th, 1919) three in my possession, which were all that remained when we were saved.

We could scarcely keep the fire going for the wood was all wet and there was nothing to burn but very light brush and what came from the wreck. However, we managed to partly dry our clothing. When night fell we crept into our hole but found that it was impossible to lay down for want of room. The reader may fancy to himself what kind of a night we had when I say we were nineteen packed into this small space. There was not even sitting room, so to give everyone a chance, we packed on top of the other – to such an extent that we could not move for the weight on our legs and bodies and dared not move for fear of losing our only shelter. Our discomfort was accentuated by cramps but to think of going out was to think of perishing. There was however, one advantage – it helped to keep us warm in some parts.

In our present position, I need scarcely say we were very glad when daylight again broke. The previous night we had divided about half the food which we had found and now went to look for more. In this we had no success as the wreck had begun to pile up on the rocks and if any had been washed up it was certainly now buried underneath. We then, or rather the captain, divided the rest of the food. After he had broken the biscuit into small pieces and handed it round, he looked at me and said, "Haven't you got any Bob?", I replied that I had not and that I never had to rush for food and hoped I never would. He then broke his in half, though it was scarcely more than an inch square, and gave me half of his. The reader may be sure that it was little, but thankfully received.

It became a matter of compulsion that we persevere in trying to find something and it was only natural to look for shellfish at low water. We also discovered a root growing in the crevices of the rocks and found it palatable.[1] We afterwards found them to be very substantial as a substitute during the whole time we were on the islands and it turned out to be of the greatest service to us. The leaf resembled that of the marshmallow but very much larger also the petal which grew in clusters, some to the height of two feet. The root itself was of a scaly nature and at every year's growth there formed a knotty substance of fiber. We had had nothing to eat except these roots and a few winkles which we were enabled to gather at low water. These we roasted, but if the reader has ever tried to fill his or her belly on them you will judge what sort of satisfaction we got. Some by this time had been eating seaweed. This was found to be very tough, so there was very little of it eaten. I don't think that I felt in the least hungry and I did then feel a tear escape my eyes. This was not very strange, for when young I was always of a tender hearted nature.

The stern of the vessel had now disappeared entirely, and the cave was now full of the wreckage. Here was one of the most tragic sights that we could possibly behold, for there jammed amongst the wreckage was one of our poor shipmates hanging by one foot as naked as the day he was born. This, as may be judged, was enough to try the stoutest hearts. Although he was close to us we could not recognize him, nor get near, as the waves were breaking high above him; but even if we could it would have been of little use, as we had no means of burying him. The poor fellow hung there all day.[2]

The weather having moderated somewhat, we now began to realize our position more fully. Right in front of us past the point of the rocks, we could see about two miles off land which we had first seen on the night of the tenth.[3]

It now became a question of trying to get to the top of the island. So after carefully scrutinizing the face of the rocks we found that owing to the abrupt nature of the rocks it would be utterly impossible to try any place except in the left hand corner. Here the loose rocks were piled up somewhat higher than elsewhere. Hipwell, Tait, Turner and a German called Fritz started to try to get up. We could not see what progress they were making for very long but, as they didn't return that night we concluded that they had met with some success. The rest of us spent the day as we had previous ones with not the slightest sign of getting anything from the wreck except firewood which was piling up. We passed the night the same as the last, except that as there were four less in the hole we were a little more comfortable.

The morning of the 13th broke the same as the others with no better prospect. The body had disappeared from the cave together with some of the wreckage. We examined the chasm high up which was about fifty feet deep with two small projections which we could bridge if we could find a plank long enough. Thus by crossing it we could find better passage than up the centre of the gully. So we searched the wreckage and took up the longest we could find; as we had no means of getting the measurement it was all chance work. This plank was six inches by three inches and about 18 feet long. We upended it on the projections on our side, then dropped it on the other. This I made use of and found it very satisfactory. It proved to be exactly what was wanted and we had a bridge which would enable us to climb the gully to the top.

*   *   *   *

The Royal New Zealand Navy had been generous in providing us with aerial photographs of the whole northern area of the island. Using the most advanced stereoscope, loaned by the Geography Department of Bishop's University, we had tried to identify the exact location of the wreck. The new map of the island (1993), sites the wreck of the *Invercauld* at the very north-west tip of the island, between the mainland and Column Rocks. This is clearly in error for the contour lines at this point merge into one solid black line indicating vertical cliffs. The accounts of the three survivors all stated that had the *Invercauld* gone in 200 yards (180m) either side of where it did there would have been no hope. From the photographs, the only spots where there would be any hope of scaling the cliffs without ropes were slightly to the south of this point. The *New Zealand Pilot* (1958) paints a fearsome picture in the official description of the west coast: "The whole coast has a most inhospitable, rugged, and iron-bound appearance. It is washed by the heavy sea thrown on to it by the prevailing westerly gales. At one place only, near the head of Port Ross where there is a narrow cleft, is ascent of these cliffs possible."[4]

More graphic is the picture painted by Armstrong of the *Amherst*: "What can I say of this coast but that I have seen nothing to surpass, or even equal, the grandeur, the savage majesty of its grim storm-beaten sea walls; standing up bold and defiant, sullenly challenging old Ocean to a trial of strength. There are portions of it where the cliffs rise perpendicularly to a height of nearly 500 feet, their sides presenting a perfectly plane surface, at their feet a small shelf of rocks, or a long low cavern; the sea breaking over the one and driving into the other with a noise as of distant thunder."[5]

We had never expected to be able to get anywhere near to the cliffs from the sea, but the second morning we awoke to a calm. Seizing the moment, we put aside our plan of heading to the top of the island, pulled up anchor and headed for the west coast. As we motored out of Port Ross we had our first seaward view of the islands and the seas sweeping through the gaps between. The brilliant sun did not hold much warmth and the wind was indeed from the

Antarctic. Grampa Holding must have had a word with the weather gods for I don't think I have ever seen a sky so blue, or sensed air so clear as in those latitudes so far from pollution. In the company of a trio of seals gambolling alongside the boat, we dropped anchor in North Harbour for a quick lunch before heading back out to sea. A tremendous sense of remoteness from the world filled us as Lance struggled to make our daily radio contact with New Zealand:

"Naw, it's hopeless Murray, I can hear you're there, but can't read you, over."

Static ...

"Sorry mate, it's no good, no good. I've got no chance of a copy, no chance of a copy, Murray."

Static ...

"I'll just listen on double four one seven and when you're coming through louder I'll get you."

The radio had provided a link with the real world, weather forecasts, chatter between other boats but now we were totally cut off. Would I ever again take the casual action of flipping on the radio or television for the very latest news and weather for granted?

Only two of the 13 of us had been to the Aucklands before so the anticipation was intense as the gnarled, warning fingers of Column Rocks rose from the sea. This scene had been described vividly in 1892 by the surveyor, Baker:

> ... a bold, bald, headland, with a rocky islet of curious conical dimensions to windward, together with a singular promontory named Black Head flanking the right. There is at the base of this promontory an extraordinary deep cavernous indentation, which were I asked to arraign before the baptismal font, I would most unhesitatingly christen the Jaws of Hell. It is altogether a striking bit of landscape scenery, well adapted, I should say, for the manufacture of legendary lore.[6]

The years of studying maps, photographs and written descriptions could only partially prepare us for the stark reality before us. I had seen the cliffs in those aerial shots, I had looked at the top sheet and seen the ominous hatch marks indicating cliffs too steep for contour lines, but the sight of those sheer rock walls rising straight out of the sea was enough to strike awe into the most courageous, even on a brilliantly sunny day with a solid deck beneath our feet. I could only imagine the terror of those seamen as their ships had been flung upon these cliffs.

The bulk of Disappointment Island loomed out of the sea haze. This ancient volcanic plug had been the site of the wreck of the *Dundonald* in 1907. Those tragic survivors knew that there was a depot on the main island but had no way of getting across the five miles (8 km) of wild water. Their story is told by one of the survivors in *Castaways of Disappointment Island*.

We had come armed with binoculars and a 400mm lens to make a permanent record – even though the chances of a successful photo with such a lens from the decks of a heaving ship would be minimal. But here we were, no more than 50 yards (46m) from the foot of the cliffs. At one point Lance calmly left the wheelhouse and went forward, leaving the wheel to gently oscillate. I stepped over the bounds when I expressed some concern and was firmly put in my place: "I can read the waters just as you read a book – so stop worrying." I did – almost – as we drew closer and closer. Sometime later I realized that it was the back swell that kept us from duplicating the fate of Grampa Holding. Paddy, with his ever present gallows humour, speculated that Grampa Holding might have made a deal to sacrifice two of his great-grandchildren, with a couple of great-great grandchildren thrown in, if he got away safely.

Using the slowest speed possible to maintain headway, we spent the afternoon working our way back and forth along that mile of coastline, reading and rereading Holding's vivid words.

We were now nearing the fatal rocks... time was so short before we struck broadside on that the only thing we thought of was, would we ever be able to land in such a place?

...It was quite plain to everyone that there was now no escape for us... It was now every man for himself with the most dismal outlook anyone could possibly imagine... Our feelings can be better imagined than described. With black rocks on three sides and a raging sea on the other, there was no chance of launching a boat.

Somewhere in one of the three coves before us, a fine ship of 888 tons had come to grief with shouts of terror, crashing of masts; right here, in these calm waters, four men and two boys had died. It was as if the ether around us held those cries for all time.

The sea is a violent creature, not only to ships and men but to the land upon which it beats. What changes had it wrought to this exposed coastline in the last 130 years? Treasure hunters have been searching for the site of the wreck of the *General Grant* and her gold since 1866. Caves collapse, great avalanches scar the face of the cliffs – could we really hope to identify the exact place? Each spot was examined in the light of the manuscript.

We were absolutely correct in saying that there was no way the ship could have gone in where it is marked on the map. Moving southward the next, smaller, indentation was a conceivable site, but it didn't quite match; the second was the most likely, for the third cove had a single rock in the middle which we are sure he would have mentioned. No way for the fourth as the cliff was too steep for exhausted, starving, shoeless men to have climbed. There was also visible at this point a rainbow streak of a rock smearing itself across part of the rock face, which I am sure he would have mentioned.

Back to the second cove for verification: right on, cave and all. During all this Paddy and the two crewmen were in the dinghy taking pictures of the *Evhoe* against the cliffs. Had we known we would be unable to get across the top of the island to the point above the wreck

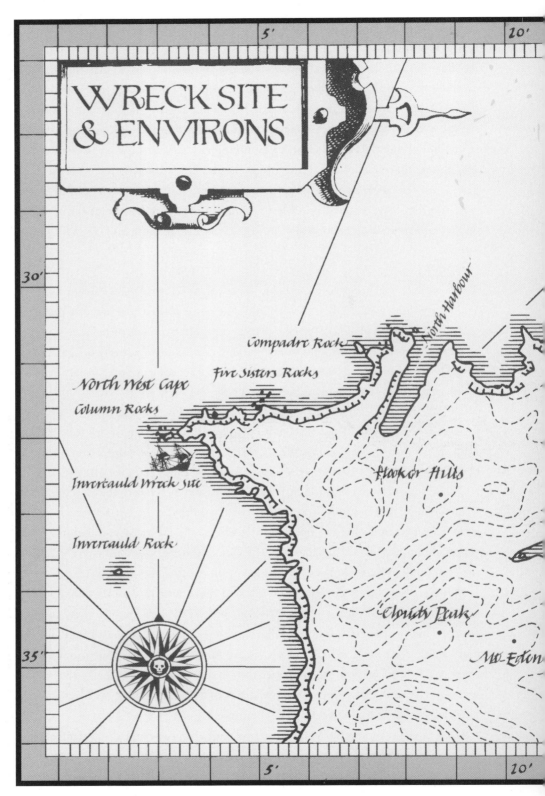

The 'Invercauld' was wrecked near North West Cape on Auckland Island, 10 May 1864.

166° 15'E   20'

Enderby Island

North East Cape

30'

~~~~~Bay

Rose Island

Sandy Bay

PORT ROSS

Friday Island   Ocean Is.

Dea's Head   Erring Island

~~ Camp

74   Shoe Island   Yule Island

~ Cove

Johnson Point

~s Cove   Hardwicke   Frenchs Island

Webling Bay

~~~~ Hill

Dundas Island

35'

Blanche Rock

Haskell Bay

166° 15'E   20'

site, I would have gone ashore – but at that time we had expected to later stand atop the cliffs and perhaps risked the climb down. My only regret was that we didn't actually land. However, we had followed the final path of the ship and we are 99 percent sure that we are able to say: "Here is where the *Invercauld* went in."

It was after six when time and tide determined that we must head back to safe harbour. There was an overwhelming sense of unreality as we opened a beer and sat on the stern bench watching Disappointment Island and the cliffs recede into the evening mist. The ever-present albatross, petrels and skewers soared high overhead. It had been one of God's perfect days.

[1]   Probably *stilbocarpa polaris*, more commonly known as MacQuarie Island cabbage.
[2]   Raynal quotes Dalgarno as having said: "We stripped them of their clothes, which had become to us a very precious thing. Having no means of burying them, we were forced to leave them where they lay; the birds of prey would soon devour their bodies." Neither the mate nor Holding mention this, despite the great detail of their accounts.
[3]   Disappointment Island.
[4]   *New Zealand Pilot*, 12th edition, (1958): p.442.
[5]   Armstrong, H., *Government Gazette, Province Of Southland*. Vol. 6, # 9, p.53.
[6]   Carrick R., *New Zealand's Lone Lands*: p.13.

10

under its lee

Castaways

During that day I heard the Captain's poor Irish Setter howling round the point for a long time. How he got there is a mystery to me. I also heard him on the thirteenth but could not see him.

Being anxious to learn what land we were on, I had been questioning the Captain on the subject. He said he didn't know but thought we might be on some island, the name which he said I have since forgotten, but which was many miles from where we were. Each of the mates were in the same mind. This did, and does now, seem strange when we know that there were two men placed on the lookout.

About noon only three of the men returned and said that they could not find Tait as he had fallen down the rocks. This of course cast a gloom over us but was partly compensated by their report that they had seen foot marks of sheep on the island. This appeared to be the very thing which I wanted to move the dormant parts of me, for I must confess that I had been thoroughly down hearted. I said, "if there are sheep on this island I will have some damper and mutton tonight."

I told the captain that I would return when I was satisfied as to our prospects and started off over the Plank. In our then state it took little preparation and it was not many minutes before I was on the way up the rocks, not knowing where I was going or what I was likely to meet. I lost some time in trying to find poor Tom Tait, but I failed entirely, chiefly through my anxiety to reach the top.

As soon as I had crossed I found the gully was covered with the wild celery previously mentioned. This was growing on very wet shallow soil and as it was necessary to pull oneself up by it, owing to the steepness of the assent it was also necessary to use the greatest caution, by not leaving go one hold before another had been secured. It was with the uttermost difficulty that I could get either a foot or hand hold which could be depended upon and many times I slipped back in doing so. However, I kept on persevering, though I knew my strength was not as it used to be. The gully opened out about one third of the way up and I found the going much better. On reaching the top, I had the first view of our position.

My first impulse was to turn and look at the place in which we had been wrecked but owing to the height it was seen only indistinctly. The gully which I had come up was only partially visible, owing to its abruptness. The water was comparatively calm. I then looked at the land which we had passed on our Port and thought it was probably two miles away and probably an island, about two and a half to three miles long; the width the other way was not to be judged by looking over it. I have long since to know this as Desolation [sic] Island as described by another shipwrecked crew, some forty years after the Invercauld of Aberdeen.[1]

The rocks, which I have before described as having seen on the Starboard side while coming in were plainly to be seen. I don't think they are more than ten feet above the surface at any time. They are very small on the surface and undoubtedly there are others under the surface which might be more dangerous. I can only say had we struck one of them it would have been all over with us unless we had managed to get some of the boats out; the success of this I am doubtful with the officers we had. If such had been the case and we got to the Leeward of that island, we might, and probably would have, found a decent landing place by keeping under the lee until morning.

I have described this as near as I can so that should anyone reading this ever get into that position, they may form some idea as to the lay of the land. Before proceeding further, I might be allowed to state that I have in my possession, and in print, accounts of eight different wrecks on these Islands, since, and including Feb. 1864, the last of which was about twelve years back on Desolation Island.

Now to go back to my trip. I found the top of the island round by the bluff covered with the muscage for some distance back, then wiry grass increasing in height as the land sloped off, then small bushes which also increased the lower it went. There was a small bay, well sheltered. On my left was a continuation of the bluff which ran to a point on the left hand of the bay. This bay had a great fascination for me as it had a good sheltered position and a likelihood of some thing to be found. It was now getting late in the day and I must be getting on. I had not gone far to the right looking round the bluff before I could see that the high bluff continued as far as I could see and curved to the right.

I may here explain that I am unable to give a statement of latitude or longitude as I did not know the lay of the island, but I don't think I would be far wrong in saying this was on the Northwest side. I could now see a large bay on my left hand and some distance off.[2] There were two small hills between the two bays,[3] therefore two gullies which left the land undulating. The hills were bare except for the long wiry grass which grew in bunches up to about eighteen inches in height. The soil was of a very boggy nature, owing to the amount of moisture from Rain and Spray. It was very bad travelling as there was little soil on the rocks near the bluffs and the foot slipped at every step. I found the foot marks and decided they were pigs instead of sheep.

It was now too late to think of going back that night, as the rain was coming down in torrents. Not far away were some rocks about seven feet high and in position to shelter me from the wind. The rain was bitterly cold and chilled me to the bone as it ran down to my boots. I might here say that I had on a good pair, they being new in Melbourne, but as I had to try to keep my feet warm by kicking the rock, either with heel or toe, and by morning they would no longer keep out the water.

Oh dear, how I wished for daylight as the rain never let up for one minute all that long night and there was not the slightest possibility of making a fire. This would kill the best man living within one week, even if well fed, so I need not say how glad I was when day began to break on the fourteenth. I did not lose much time in going back as I had seen all that could be seen of consequence without getting down to one of the bays. This was now my sole object.

The wind and rain having somewhat abated, I made fairly good progress going back. It did not take long as the distance could not have been more than three quarters of a mile, so I was there early. I found that it was as dangerous to go down the slippery Gully, as to go up, perhaps more so. However I got down without accident.

I need scarcely say the party at the wreck was very pleased to see me back and hear what news I had to impart. I had proved pigs on the island, had a very rough night, and seen two fine bays. I thought I had done very well for the time I had been up. I was now told of three other men having gone up. These I had not seen. All the others were now down. Poor Tait had found his way back and was in a very bad way, much bruised up and quite delirious, calling out for tea and tobacco, neither of which we could supply. He had hidden himself under some bushes, which was why no one had found him. No more food had been found and all they had had were some winkles which they had been able to gather at low tide. I don't think the tide rose and fell more than five feet at any time. The wreck was still piled up as I had seen it.

It will be seen that our prospects here were

a perfect blank and so I placed the following proposition before them as to future movements; that the first thing in the morning we leave one man with Tait and all the rest would go up to the top, where we would at least find some shelter, which was impossible here. How were we to get up? Well, it was not too bad with care. We would send two men down to the small bay to explore it and they would then come back to meet the rest of us. Once we had met up with the three men who had already gone up we would make for the large bay where I felt we would find something to our advantage. The proposition appeared to give entire satisfaction, especially as I said I believed we would be able to find some sealions about the bays. The night passed just about as the others had done.

We were early astir after another long night and took another look round the Wreck without result. Having no packing or other preparations to make, it was only a short time until we were underway leaving Tom Turner (voluntarily) to look after Tait. We told him some of us would return to acquaint him with our progress. If Tait recovered he was to take him up.

* * * *

Smith recounts this leave-taking: "One of our number had the kindness to volunteer to stop with him until death should put an end to his sufferings, for he was so much bruised that we were all sure it would not be long before he died. It was heart-rending to think that we could give him no help, while the same fate was staring us all in the face. We shook hands with him before we left, and bade him a mournful good-bye."[4]

We reached the plank all right, but here some difficulty arose as some were too frightened to run the risk of crossing without assistance. We had some mufflers and tied them together 'round the waist of the most timorous. They would have been of little use, had anyone fallen, but they gave a little confidence. In this way we got everyone over. It was quite a job to get them up but we managed by getting behind and giving a push up occasionally. All managed to cross without mishap.

We had only just reached the top when we met the three others coming down. To our great surprise, and pleasure, they had with them the best part of a small pig, which they had captured by running it down and falling on it. They were going to take that down to us.

They had, they said, eaten the liver raw while still hot as they had no matches with them.

We were so taken up with the unexpected prize that we hurried down the hill to where we could get some small brush to make a fire and it is needless to say how soon that pig had vanished for I don't think it weighed more than twelve pounds. This was, however, encouraging and confirmed my statement of there being pigs on the island. After my appointing Big Dutch Peter and Fritz to go down to the small bay I pointed out to them where they could come to us. I intended to take all the rest round towards the big bay, which was then partially visible from where we stood. They could not have had more than a quarter of a mile to go.

For some reason or other Turner had left Tait and had followed us up to the top. He stated that Tait was in a very bad way and that he had left him there. It did seem that he was too frightened to stay with him. It was difficult to get anyone to do anything at that time.

We then climbed the hill to obtain better travelling and had gone only a short distance before Peter and Fritz caught up with us saying that they had been to the small bay and found nothing there. I knew they had not had time to go there but could say little then.

Turner, Hipwell, Cook and two others decided to go along the bluff to see if they could catch another pig while I took the rest down the ridge of the next hill to the large Bay. This hill was the last before reaching the large bay and ran parallel with it, as well as being the longest running towards the southeast, thus shortening the distance down to the bay which could not have been more than half a mile distant. We passed the rocks where I had spent such a miserable night and were going along the ridge of the hill when we caught sight of a pile of stones to the left of us, and from the appearance of them we were quite satisfied that they had been placed there by man.

We might have gone down about half mile when it commenced to rain (and when it rains there it does rain some), so we made for the first and only apparent shelter. This was on the side of the hill towards the large Bay. Our speed, though it was never very great, was greatly reduced by our wet clothing. It was at all times difficult to get some of them along. All who had gone to look for another pig had caught up with us except for the Cook. He had, while we were in deadly peril and pulling our arms out of their sockets, gone to his berth and put his best clothes on over his others. Now they were soaking wet and were a great encumbrance to him and in fact it was not without great difficulty that he was able to get along.

We soon reached the small sized brush, which might have been eight feet high and here they would go no further. So there was no alternative but to stay there, though if we had gone down about two hundred yards further we'd have found decent shelter. After wasting far too many of our valuable matches we got a fire started. This was much needed as the long grass had wet all up to the waist and the rain had completed the rest. The rain was now coming down in torrents and cold with it. We must remember that it was then nearing the shortest day. As it was nearly dark we had been expecting the cook to join us for we had seen him struggling towards us in the long grass. We knew he had seen us as well as the smoke for he had answered our call but he did not show up that night. It would have been a plucky individual who would have gone to his assistance through the cold rain of a night like that while wet to the skin. But even so we did not know his perilous position.

Most of us lay down but I am sure that none of us slept that night and how we passed this night the reader must judge for himself. All I can say will not explain it for it rained all night as it can rain there. With our combined efforts we could not keep the fire alight, having only green wood which we cut with our knives or broke down and had to give it up entirely before morning. When I look back on these times I often wonder how we stood the hardships so long.

Were we glad when day broke, well rather. The rain had ceased and there was nothing to keep us here. We had eaten some of the roots, which were so plentiful here and of a somewhat better quality. Some had actually tried to eat grass but had found it tough and of rank flavour.

Our first thoughts in the morning were for the cook who could not have been more than a quarter of a mile from us; although it was clear of bush we could not see him and we all concluded that he could not have lived through that night of misery. I choose two of the apparently strongest to go to look for him. My plan, which I communicated to them, was to go up the hill to where we had branched down, then for the best travelling continue down the ridge and on to the large Bay.

Thus we parted and on looking over the next hill we could see quite plainly a small pile of stones in a conspicuous place which I had noticed before. This the captain pronounced to be a Cairn.[5]

It will be realized that by this time we were becoming very weak for want of food and some were becoming careless as to their future. In consequence it looked as though my plans would collapse at any moment. I did however, manage to get them to make one more move.

We went up the hill a little way to secure better walking, then along the top to a point where the grass terminated and the bush began.

We had not gone far when the two men who had gone to look for the cook caught up with us, saying that they could see no trace of him. While I knew that I could go straight to the spot where he had last been seen, my duty lay with the largest party.

[1]   The *Dundonald*, 1907.
[2]   North Harbour.
[3]   Laurie Harbour.
[4]   Smith, Andrew, *The Castaways*: p.11.
[5]   No sign of cairns were found. Similar known ones on Adams Island have fallen due to wind and weather.

ALL COLOUR PHOTOGRAPHS BY PADDY RYAN, EXCEPT P.103 (TOP & LOWER RIGHT) AND P.104 (TOP RIGHT & LOWER): MADELENE ALLEN.

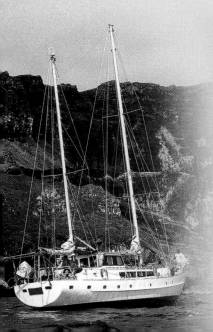

AUCKLAND ISLANDS
NATURE RESERVE
ENTRY BY PERMIT ONLY

(*Page 97*)(*Top*) Cape pigeon; (*Middle*) North West Cape, near site of *Invercauld* shipwreck; (*Lower Left*) Crab was an infrequent treat for the survivors; (*Lower Right*) Kelp, *Durvillaea antarctica*. The Auckland Islands have the world's richest kelp beds. (*Opposite page*)(*Top*) North West Cape; (*Lower Left*) Madelene Allen's 1993 expedition sailed to the Auckland Islands on the *Evohe*; (*Lower Right*) 1993 Expedition members.
(*This page*)(*Top*) The *Evohe* at the *Invercauld* wreck site; (*Left*) Dense ground cover on difficult terrain prevents easy movement across Auckland Island.

(*Opposite page, Top and this page, Top*)
Bleak landscape of the Auckland
Islands added to the shipwrecked
sailors' misery; (*Lower*) Expedition
members trudging through tussock
on a typical Auckland Islands day,
December 1993.

(*Opposite page*) The stream Robert Holding crossed when leaving the site of the Hardwicke settlement on his explorations; (*Left, Top*) Author examining chimney bricks at Hardwicke; (*Middle*) The *Amherst* spar; (*Lower*) Remains of German Transit of Venus expedition plinth; (*Lower Right*) Second mate John Mahoney's headstone at Hardwicke.

(*Above Left*) Castaway depot; (*Right*) Author with family and other 1993 Expedition members, from left: Jim Attwood (yellow leggings), Bruce Allen, Brenda Allen, Pete McClelland, Madelene Allen, Robin Allen. (*Lower*) Remains of the *Grafton* at Carnley Harbour. Survivors of the *Invercauld* and *Grafton* wrecks never knew they were sharing the island, although on one occasion sailors from the *Grafton* thought they saw smoke.

11

through the bush

Took Sights

*"The generally cold temperatures, high rainfall and frequent cloud
make fieldwork difficult and necessitates extra caution on inland trips."*

Department of Conservation Management Plan

ONE of our major projects while on the island was to climb to the ridge of the Hooker Hills and cross overland to the top of the cliffs above the wreck site. We had set aside three days for this venture, one day to get up and hike part way across, and two days to go to the site and return, which would give us time to find all the landmarks which Holding had mentioned. Camping on the island is not usually allowed, but DoC had agreed that this was a worthwhile project.

We laid out the map and compared the tightly-spaced contour lines with the view before us – a true lesson in visual topography. A heavily wooded slope rising steeply from the sea to 900 feet (275m) in just a kilometer, then a more gentle rise to the top of another 450 feet (140m) in about the same distance. The base rock of these hills has a deep covering layer of one to six feet (1.8m) of peat. A falling tree, or instability caused by the heavy moisture, can result in a full earthslide which takes everything in its path. Facing us were three parallel slips which began near the summit and continued about two-thirds the way down to the sea. Although steep, they would provide an easier path. We took a visual angle to these, for compasses are totally useless in this area due to magnetic anomalies (variations) and we knew they would be invisible once we got into the bush. John Hay, the district surveyor, had reported in 1891 that "from two different points in the locality of Port Ross I observed bearings to the ship with prismatic compass from a distance of two to three miles (3.2-4.8 km). In one case my bearing was in error 44° westerly in the other case the error was 175° westerly."[1]

Jim's tape picked up Lance's instructions before we went ashore:

"...everyone has to scrub up, every time, no matter how clean they think their boots are. When you're going ashore with anything that could possibly contain a mouse make sure there is no mouse. I will be sure when you've taken your pack up to the forward cabin, emptied the whole thing out and repacked. I'm sorry, but that's just the name of the game. It's got to be done every time. I don't want any slip-ups going from this boat. It's very, very important.

"At all times I want to know where everybody is. When you're going to shore you never, ever, get out of the dinghy without telling the driver what time you want to be picked up. If anybody wants a lifejacket, all you have to do is ask. As far as where you go on shore is up to Peter, but when it's safe to put you ashore – we'll make the decision. When it's time to get out of the dinghy, you'll be told by the driver. Go only when you're confident that you can do it yourself. I don't want anybody playing he-man here, showing us how brave you are. When you're in the dinghy, you have to realize that you're a long way from home and an accident is not just a pain in the butt to you, it's a pain in the butt to everyone else. So be careful. I think that's about it. Pretty much commonsense."

We were put ashore on a narrow ledge of black volcanic rocks on the north shore of Laurie Harbour, just opposite Wilkinson Point. I have never seen anything like the wall of vegetation which faced us; the word 'impenetrable' now has a completely different meaning. The only way I can think of explaining its thickness to a Canadian is to ask the reader to conjure up an image of the thickest alder scrub along a northern river that you had ever had the misfortune to scramble through – now double that and factor in branches that will not bend. We fought our way upward, stepping over, crawling under, often losing sight of the person just four or five feet (1.5m) ahead. My husband had his sleeping pad tied to the outside of his pack and within minutes it was torn to shreds. We thought we might go back with the tale that he had thrown it at a charging sealion.

The rata forest seemed to go on forever. Raynal paints a graphic picture:

… a most remarkable kind of iron wood, with a thin bark, whose trunk measures from ten to twelve inches in diameter. This trunk is generally twisted in the most fantastic fashion; a condition which may be attributed to the constant struggle it has to maintain against the winds. It seems that in the moments of respite it hastens to resume its normal mode of growth, and to rise perpendicularly; then comes the buffeting wind again, and beaten down anew, it bows, and writhes and humiliates itself, to shoot aloft once more for a foot or so, until soon it falls back vanquished, and is bent towards the ground.

Sometimes these trees, being wholly unsuccessful in their attempts to rise erect, crawl, as it were, along the earth, disappearing every now and then under hillocks of verdant turf, while the portions visible are coated with mosses of every description. The thick gnarled branches share the same fate as the trunk; they attempt at first, as it does, to spring towards the sky, then forced to abandon their aspiration, they take a horizontal direction.

They bear, nevertheless, a thick foliage which shelters, as a roof might do, a whole subordinate world of shrubs, heaths, and marshy plants.[2]

He adds a footnote at this point from Dante's *Inferno* (Book i, Canto xiii 2-7):

We entered on a forest, where no track
 Of steps had worn a way. Not verdant there
 The foliage, but of dusky hue; not light
 The boughs and tapering, but with knares deformed
 And matted thick: fruits there were none, but thorns
 Instead, with venom filled.

Passing this way in 1907, Charles Eyre, a survivor of the *Dundonald*, described their crossing of the island:

It was now an awful forest that we now came to. The trees were nearly twenty feet high, and you will understand that they sloped down from the spot where we were towards the sea. And now I fear that some of my readers may not believe me, but I assure you that I am telling you the truth. That forest was so thick that we could not walk through it – we absolutely walked on top of the trees! So thick, so dense, and so interwoven were the branches and creepers, that although they gave beneath the feet like a spring mattress, we could walk – or more often roll – over them without falling through.

 Not but what we did not fall through again and again, for when rolling down we would come to a break in the bush – either gaps or places where the foliage was not so thick – then, with a crash, a rending of branches, and tearing of clothes, down we would go slipping to the ground beneath, and have to clamber up again, helping one another as best we could. By the time we came to the end of the valley our clothes were literally torn off our bodies.[3]

Despite strenuous training in the months leading up to the expedition, I realized from the outset that my 50-year-old body, laden with camera equipment, was not up to toting a full pack through this terrain. It was a humbling experience to have to accept 'Mr P's' offer of help. Without this young man's assistance, I would never have made it to the tops.

 Long after we felt we should have struck the slip we were still struggling through bush. Being packless, I was nominated to climb a tree to try to spot the break in the bush. It had been sometime since I was last found in the branches of anything, let alone a rata tree in a dense sub-Antarctic bush, but mission was accomplished. From six or seven feet up it was just possible to see a slight clearing which had started well to our left and below us. We had been struggling unnecessarily for quite some distance.

Cutting eastward, we finally reached the slip and collapsed to get our breath. Crouching beside a stream which rattled over the rocks, we slurped handfuls of clear, dark water. Despite the tannin flavouring from the peat, it tasted wonderful. After we had all drunk our fill, Pete drew our attention to clusters of tiny white worms hanging onto the rocks under the water. It seems to have done none of us any harm for no one contracted the sub-Antarctic equivalent of 'beaver fever', if there is such a thing.

We scrambled upwards through mud and gravel, grateful for the relatively unimpeded climb which didn't last nearly long enough. The line where the slide had started was as sharp and smooth as if it had been drawn with a straight edge. Waiting for us was an even denser, finer bush which Clifton described as "an impenetrable tangle through which a passage can only be chopped inch by inch with a slasher".[4] I must add that we did not use a slasher but wriggled our way through in whatever direction we could. Clifton goes on to explain, most appropriately, that "altitude is perhaps a misleading term, for the height at which the bush grows depends on the amount of shelter from the wind."[5]

We finally broke out of the scrub and into the great thick tussocks of *poa* grass which cover the treeless tops. These tussocks, several feet across, rise to shoulder height from the knee-deep bog between. We had the choice of either slogging through the peat, weaving between these great chunks of vegetation or leaping, stretching and straining to get from one tussock top to the next. I found it very rugged going but those with full packs had an incredibly difficult time for the wind was so strong that it was almost impossible to maintain balance even had the ground been level.

A bleaker landscape would be hard to imagine; fingers of craggy rock loomed out of the mist that swirled and enveloped us; rain streamed down our faces as we canted against the howling wind, which was so strong that it was impossible to hold a video camera steady to the eye. I wondered how the *Invercauld* survivors ever coped with the physical and mental demands of this island.

One description of Holding's that we had found hard to credit was that of his walking on top of the bushes. It was not long before we came to realize that he had not exaggerated. Myrsine (*Myrsine divaricata*), a thorn-like bush, was knitted together so tightly that we were often walking on a second level. At one point we were confronted by a short rock face with a drop of six or seven feet (2m). Tight against this face, blocking the drop, was huge myrsine growing up from below. Taking a leaf from his great-great-grandfather's book, Bruce showed us the way. With full pack he launched himself onto the top of the bush and rolled over and over until he dropped off 10 feet (3m) away. We followed his example with varying degrees of success. Paddy's shoelace hooked on to a branch and he had to be cut free. My first roll was successful, but on the second my shoulders hit a weak point and I pitched backwards, head down at a 30° angle where I hung until rescued by my brother. It was certainly not an area in which to venture alone.

We struggled onward, finally taking refuge behind a barren ridge of black rock which protruded from the bog and leaned sharply towards the sea. Tiny flowers bloomed in its sheltered crannies. Grampa Holding's remark: "When it rains there it does rain some!" was well remembered and quoted as we ate our rather battered, soggy sandwiches.

We were awestruck by the absolute starkness and desolation around us. How had the crew of the *Invercauld*, at the beginning of a vicious southern winter with no food, few matches, many of them shoeless and all poorly clothed, managed to keep up hope? Indeed, many of them didn't. Holding, coming into the voyage in good physical condition and with a knowledge of the land and experience of rough living, had a much greater chance than many of his comrades; is it any wonder that they turned to the one man who could lead them?

For the first time in my life, huddled against the violence of a piercing wind which drove the rain before it, I understood how men could simply give up the will to live and lie down and die. It was in this area, in weather exactly like this, where the cook, exhausted and burdened by his wet clothes, had lain down for the last time. As we stood looking over the swaying tussocks, it was clear that a search for a prone body would be successful only by actually walking to the exact spot. It was not far from here that young Lansfield had gone to sleep in the snow – and miraculously had awakened – only to greet another hopeless day.

By two o'clock the storm showed no sign of slackening and the mist, if anything, had grown thicker. The decision had to be made whether to struggle on to try to reach our goal, (not knowing what the weather would be like the next day) or to admit defeat and return to the ship with the day party. Thus far in our two hours on the tops we had found nothing remotely resembling a place where we could set up a tent. Everywhere was bog, stream or tussock. If we were going to continue, we should get another hour or two's hiking in; on the other hand, it would be foolish to continue for another two hours and then be caught by darkness with no place to sleep.

It was ironic: the *Invercauld* survivors had no choice but to cope with their misery (or not, as the individual case may have been). Here we were, in early summer,[6] equipped with products of the most modern technology in outdoor living (Goretex jackets, lightweight tents and sleeping bags, portable stoves with fuel bottles weighing only ounces, freeze-dried foods, and a radio to the ship at our backs), questioning the wisdom of spending the night on the tops.

There was no point in all of us slogging around searching for a campsite and risking someone getting separated from the group in the fog so it was decided that Pete and Brenda, the most skilled in outdoor survival, would search. We gave them an hour, which would give time for the shore party to get back down before dark should they be successful. Somewhat tongue-in-cheek, Pete had observed that all we had to find was two albatross nests close together. As albatrosses return to the same area year after year the ground, for a radius of perhaps 12 feet (3.5m), becomes flat and hard.

We took what shelter we could beside a rock overhanging a stream bed filled with muddy red water. The wind and cold were cruel and poor Eric, who didn't have proper gear, truly suffered as he helped Paddy with his equipment as they clambered over the rocks. The enthusiasm of the *Evohe's* crew to get involved and to help wherever and whenever they could was wonderful.

I confess to having had very mixed feelings when Pete and Brenda reappeared through the mist to report that they found nothing remotely resembling a tent site. We had two choices, either to simply prepare ourselves for a miserable night in the bog and then not know if the weather would clear enough to continue the next day, or we could go back down. The chances of fine, or even decent weather the next day were not good. (The coastwatchers in 1941 had recorded only 44 days in the whole year when the cloud base was above 660m (2165 feet) and on 124 days it was below 300m (985 feet.)

The decision had in essence been made before they set out and it was a 'no go'. I felt somewhat ashamed of our weakness, considering the history we were following, and yet I knew that we were being sensible. It was disappointing to admit defeat after all the work to obtain permission to camp and we felt somewhat sad to miss the experience of spending a night on the tops. However, we had been up, we had seen the type of terrain which the *Invercauld* survivors had faced. More importantly, we now understood, as we never could have if we had not been here in abysmal weather, the sailors' struggle to keep going.

If anything, the hike back to sea level was more difficult than the struggle upwards. At last we hit the slip, although I am not sure which one it was. We slithered down through the mud, in high enough spirits to have sung a line or two from Flanders and Swan:

Mud, mud, glorious mud.
Nothing quite like it for cooling the blood,
So follow me, follow,
Down to the hollow,
And there we shall wallow in
Glorious mud![7]

My gaiters will have Auckland Island mud on them forever.

Spirits never flagged in our group, thanks in a large part to Paddy's continuous patter of dreadful jokes and friendly banter between Brenda and Pete. Leaving the slip behind, we followed the stream down a deep gully through the dense bush. Giant tree ferns, moss hanging in great garlands: it was a botanist's paradise. The hills provided shelter from the ferocious winds, and here a more fragile, rainforest-type vegetation flourished, a tiny micro-climatic region reminiscent of New Zealand's Fiordland National Park. It was here that we had the only slight accident of the trip. I was having difficulty negotiating the steep downward slope with my bifocals and had taken them off so that I could keep my feet in focus. Being used to the protection which glasses offer I turned my head quickly, without thinking, and rammed a branch into my eye. In pain and without binocular vision the rest of the day was a nightmare.

Finally we reached the fog-enshrouded shoreline. It was 6:30pm and our radio check-in was not until nine. The thought of waiting for two-and-a-half hours in the clinging mist which hid the ship from view was not appealing. Pete had brought his rifle in case we came across any wild pigs or goats (which were being eradicated) so, not even thinking of trying to raise the *Evohe* on the radio, Pete fired two or three shots which echoed again and again against the hills. One wag shouted "Oh my god! He's hit a sealion!" We learned later that Lance had kept the radio on all day in case of emergency and we could have called in. How often the most obvious solution is the last to be considered.

For the seafood lovers the thought of all those delectable mussels lying in the water was too much, so Paddy waded beyond the tideline and gathered dinner. Before long we heard the throaty rumble of the dinghy's 15 horsepower motor as it crept along the shoreline looking for us. We shouted and Brenda waved her bright red jacket; the motor opened up and we were soon back on board. Our previous care with the water supply was rewarded for we were allowed a quick, hot shower. Crouching in a tub no larger than a foot bath, I felt the thin stream of water sliding over my body was heaven.

That evening, nursing our drinks, we talked long into the night about our experiences, our feelings and impressions of this wild and wonderful land, unable to forget the sufferings of those who had drawn us to these islands.

[1]   Carrick R., *New Zealand's Lone Lands*: p.71.
[2]   Raynal, F.E., *Wrecked on a Reef: or, Twenty Months among the Auckland Isles*: p.88.
[3]   Escott-Inman, H., *The Castaways of Disappointment Island*: p.237-238.
[4]   Clifton, L., *Auckland Islands* (1946): p.17.
[5]   ibid.: p.17.
[6]   December.
[7]   Flanders & Swan, *The Hippopotamus*.

12

Raynal

cliffs

We passed down the ridge until we reached the brush which much resembled that in which we had spent the last night. They made up their minds not to go any further and it was impossible to get the party further that day so we made camp on a hill which sloped down to the large bay [Laurie Harbour] on the right and towards the centre of the bay [North Harbour] which was plainly to be seen therefrom. The next hill [Mt Eden] from the large bay was the one on which the cairn was placed and evidently was intended as a land mark in coming up the bay.

We had a feed of roots which were much more plentiful here; they were of a sweetish flavour but made us feel thirsty. We had divided ourselves into two parties to be a little more comfortable under such circumstances – that is if anything could be called comfortable. We then started to make some sort of shelter of some small wood, perhaps nine feet high and two inches in diameter. By cutting and breaking it we soon had some to lay on to keep us off the wet ground. We stuck some of the wattles on end and tied them together at the top to shoot off some of the rain. Some of the men would not even do this, but lay down anywhere and the consequence was that even here, there was not enough room for comfort. They that had not tried to make cover were the first to take what others had provided and would roll on top of anyone. We laid down near the fire which we had managed to light.

I don't think anyone slept again that night as it took all our time to keep the fire going. By that time we were getting used to sleepless nights. Our matches were becoming scarce and we had to preserve them now to the uttermost as we did not know how long we might require them. Thus passed the night of the fifteenth of May.

Needless to say we were on the move at the peep of day as there was nothing here to induce us to stay. I tried to get some of them to help me cut a road through the bush, which was so thick just there that it could not be seen how far it extended. There were only two besides myself who would try and even they too gave it up in a few minutes. This was disheartening to me. I may here say any man could have got through without trouble but I thought by cutting out the brush they would take heart and push on to the bay. Instead they passed the day eating the roots and drinking water. We were never short of water as little further up the hill there was plenty in puddles amongst the grass. Most of them were too lazy to even go for that and as the poor boys Liddle and Lansfield had been spared up to that time they were sent to fetch water in boots and so'westers.

I then proposed to take all who were able to go with me down to the large bay, breaking down branches as we went as a guide for the return of one, whom I would send back to fetch the remainder as soon as possible. We had, up to this time, been passing along parallel with the head waters of the Bay and opposite our present location it had opened out nicely.

Finding it impossible to get them to do anything for themselves I became disgusted. They would not try to go more than a few yards through the wattles – here loosing the best and only chance of saving their lives. Seeing at last that they would not try, I gave it up myself and decided it was useless to try to do anything for them. Those who would not move were a detriment to all. The fatal results will appear as the reader progresses. Needless to say, we were thoroughly downhearted and most very morose by this time with the consequence that little conversation passed between us. It is probable that had we been better acquainted with each other things might have been somewhat different.

As it was, the day passed almost in silence and the following night was similar to the previous one except that we had a sprinkling of snow with a little frost. Someone sent young Lansfield to fetch water which lay in puddles along the ridge. When it became dark and he had not returned I became anxious for his welfare and knowing he had only forty or fifty yards to go, feared something serious had happened to him as there was no bush to go through, nor any other obstacle to contend with. Expecting to find him easily, I started off to look for him but after going a long way beyond where he needed to have gone and calling out every few yards, I had to give it up. Frequently in the night I called out in the same way, still there were no signs of him until daylight when he returned. After questioning him I learned that he had lain down and gone to sleep. There was then about an inch of snow on the ground and it is a mystery to me what was his object in doing so, or how he pulled through that night. Poor boy must have suffered terribly.

There was little difference in our prospects and no chance of getting them from that place. The roots here were of a better quality than we had been able to procure previously and they had as much as they wanted. We had also found quite a variety of small berries which grew on small vines on the ground which were quite palatable. There were others which grew on a low bush found between the border land of the long grass and the bush proper. The berries, all about the size of red currants, were of three kinds or colours, red, white and bluish. The latter were not eatable through their rank flavour [myrcine or coprosma berries]. This may seem strange, as one bush could not be distinguished from the other.

The fate of all hands was now hanging in the balance, but they did not recognize it, or if they did, they were too careless of their lives to move. So after weighing things up and knowing that nothing would move them I told them I intended going back to the wreck to see if anything had been washed up. The Boatswain said he would go too. He was perhaps ten years my senior and in pretty good condition so I thought he might be agreeable company. The next morning (May 18th) we lost no time in getting away, nor on the way, except to try to find the cook. I was positive of the place where we had last seen him and we soon found him near some low rocks lying on his back, quite dead. He had evidently died the same night he had reached there, having apparently sat down and then laid down. There was but little grass just there, so he had no shelter whatever as the ground sloped off very abruptly. We could see quite plainly that he had slipped and giffiled about ten feet. We did all we could for him, which was to cover him with grass. The reader will understand what our feelings would be at such a time without my enlarging upon it any further.

We reached the Wreck in good time and had a good look around. Poor Tait lay where we had left him, of course he was quite dead which did not increase our love for the place. All that we could do was carry him out of the old camp and put him under a rock and cover him with bush. Having anticipated this, it was not nearly so heartbreaking to us as it may have been had we expected to find him alive. Still it was bad enough not knowing how soon either of us might be in the same position.

There was little wind and the sea was reasonably calm in consequence. The place looked in a terrible state. There was, of course, some little wreckage there but nothing like what had been piled up. What was left was little better than splinters piled about five feet high. On searching among this we found a few pieces of what looked like meat, but of what kind it was impossible to say.

We soon had a fire, cooked some winkles by roasting and having picked out all the splinters we could, tried to roast the pieces of meat. It was too rotten to hold on a stick and was difficult to eat. The rest can be imagined. Even now when I think of it it gives me the horrors. With darkness coming on we made ourselves as comfortable as possible but did not sleep much as the seas were breaking and the spray was flying over us. Needless to say the night was a miserable one.

We were on the move early, glad to see daylight again. There was little more to be found, so we got some roots and a few winkles to eat. About midday four others joined us and they also had a good look around to try to find something. Under a large rock someone saw the pig which had been on board. We lost no time in trying to release it, but it was jammed so tight that we had to get a plank as a lever to prise up the rock. I took it by the hind legs. Guess my surprise when it came in two across the loins. I figured it at about 120 lbs. but ROTTEN I should think. The reader will perhaps say, "I would not of eaten that," but circumstances alter the cases. Please consider our position, then I think you will see that we had to eat anything we could obtain or die where we were. Did we eat it? Of course, although even yet I shudder at the thought. I don't wish to dwell upon that subject, it is too abhorrent.

I have mentioned a rock about six feet high with a top about four feet square. It appeared to have slipped from up above and landed on its end, apparently bound to the cliffs on the two sides, the other being clear. Casting my eyes around the face of the cliff I saw a large bird sitting on top of it. I thought what a meal we could have if he could be secured. It did not take me long to come to a decision. The only way to get at it was by crossing the plank then to crawl along the sloping surface the best way I could. So without more ado I started off without telling anyone of my intentions. On getting across I found the bank very steep – somewhere near seventy degrees. I had some sixty yards to go to the rock, all the way very dangerous on account of the slippery and rotten nature of the soil.

On reaching the rock, I found it higher than my head. On the corner there was a small projection about three inches square at an angle of about forty five degrees. The footing was very dangerous. It was impossible for me to get more than three fingers on the top of the rock with the bird sitting there. A ticklish job at best, for the rocks were quite one hundred feet straight below and if I slipped nothing could save me. After taking a look at the distance I would have to fall (this gave me the shivers) I made up my mind to try at any risk. I now took a look to see if anyone had noticed me so that if I fell they would be able to tell what became of me.

There was but one way to get up the rock and that was to put the toes of my left foot on the projection and try to pull myself up by my fingers until I could get my left elbow on top. This was what I did so that the left elbow and the left toes were the only support I had. As soon as my head was above the rock I saw that the bird had a string of wild celery near its legs. This made the capture so much more difficult as the bird was too tall for me to think of catching it by the neck so this had to be done by the feet, if at all. We had a good look at each other and I easily recognized it as of the Albatross type, a very large bird which was known to put up a stiff fight, at which I was seldom backward in those days. I could certainly not hang there long, so my mind was soon made up. If I was to capture it, it was to be done quickly. The idea I had formed was to grab it by one leg, put that in my mouth and trust to the rest to get down. This I tried to do, but to my great surprise, I grabbed hold of a stem of the celery instead of the bird's leg. Perhaps it was just as well that it was so, for I don't think it would have been possible for me to have escaped being killed had it been otherwise. As it was, the bird must have got a fright too for it was not there when I had time to look at the rock again. I then went back to the others and told them of my adventure and I don't think they ever believed me. That didn't worry me, but losing the bird did for a long time.

*   *   *   *

Back on the tops, Smith describes in simple terms what was going on while the others were at the wreck:

> We camped here for some days in hopes that we would see some ship passing. We kept up a large fire to keep ourselves warm and as a signal should any ship come in sight. But day after day, and night after night passed wearily on, and no appearance of any ship coming to our rescue. We were very uncomfortable in this place, owing to its exposed position, while there were only a few bushes to shelter us from the heavy showers of hail and rain that were continually falling, and the wind was still blowing very hard.[1]

On board the *Evohe*, I expressed surprise that they would stay for so long in an exposed position. "Why, in heaven's name," I asked, "wouldn't they have immediately moved down to the coast?" Mickey, who had been at sea all his life, explained that seamen don't like the coast but wanted either to be far at sea or well inland. The heights would have given the survivors confidence and hope. Once lower down they no longer had a vista of the sea and their fear of missing a passing ship would have been enough for them to risk starvation and exposure on the tops.

The dense, unknown forests of the lower land held greater terror.

The days passed on in similar manner with little change. No relief, no excitement to liven us up and nothing of importance to record. To try to follow dates might be misleading as we had no way of keeping track and little interest in one day over another so we lost track. I shall therefore in future, almost entirely ignore them. To repeat our troubles day by day would be neither instructive nor educative, however there are two events which I have good cause to remember. I pass on to the 30th day of May.

During our conversation the Boatswain suggested that as we had no food nor likelihood of getting any, we draw lots. This, or something worse, I had expected for some time. So, to cut matters short, I said: "No, Boatswain. If I have to die it will be while looking for food and not while eating another man."

It took me but a few minutes to decide upon my future plans. In fact, I may almost say they had already formed. I said: "Boatswain, in the morning I am going to start off to the small bay. Will you come with me?" He said he would, so that was settled for the night at least. With this resolve we got into our shelter, but not to sleep for me, for being of a suspicious nature I did feel and dread treach-ery but had no misgivings while awake. I knew that the only one who the Boatswain feared was myself and thought probably I might be the chosen one to start up-on. The night passed off quietly however, and as soon as daylight appeared I was on the move with a full deter-mination to get away from there as soon as possible.

When, however, we had begun to move, Big Dutch Peter said he would go too. I said: "No, Peter, I don't want you, for you have deceived me once. I am going to the Small Bay – you deceived me by telling me you had been there, while I know you never went." The Boatswain then said he would not go. This appeared to confirm my sus-picion of treachery and I would have had but a poor chance with the two of them, as a blow on the head unawares would have given them a great advantage.

As things turned out, it was perhaps the best for me. I took a last look at the wreck and bid them good-bye. I did, however, expect them to follow me up and go to where I had taken the others. I may now state definitely that these five were never seen or heard of after I left them. Therefore the Reader must judge for himself as to their fate.[2]

Thus the first day of June found me up on the island again in search of food and adventure. I don't think it was more than half mile from the top of the island to the Bay but the wet bush made it very bad travelling for the scrub became rather thick in the gully leading to the head of the bay. The first thing which I noticed was a long flaggy grass about two feet high. To my great surprise amongst it I found the skeletons of two camp frames, one not more than eight feet by ten and the other smaller still. These looked as though they had been used with tent coverings. In one, still standing, was a stretcher about eighteen inches wide. This was, of course, a really agreeable surprise, showing without a doubt that someone had been there and gave hopes of their return at some future time.

Being anxious to explore further I did not spend much time with the "House Remnants" and, as the water was not more than a few yards away, went to this without delay. I there found the one thing which I was particularly looking for – a nice, well-sheltered sand beach. It had a rocky ridge along on the side rising as it passed to seaward. It being low water, I could see that the beach ran all down the left hand side of the bay while the rocks which formed the right hand side gradually rose as they extended out towards the sea.

On coming over the sand beach I saw, some distance down, two seals moving about on the sand. This was another agreeable surprise for this was what I had been looking for before anything else. Being by myself and not knowing their habits I thought it best to leave them alone. I turned my attention to the rocks for shellfish and found the limpets very plentiful. This was certainly encouraging.

I might here explain that to the uninitiated that the limpet has only one shell, very much like a sou'wester in shape. They are very easily removed from the rocks on which they cling if a knife is thrust under them quickly, but if they are touched only lightly they need to be cut from the rocks. I was then quite ignorant of this and had to learn by experience. They were of two colours, some light yellow while others were almost of a liver colour, doubtless owing to the sex. As soon as I found them I tried to eat them cold, and of course, raw.

While doing this my thoughts were of the Captain and the others who I had left behind in the hills. In hopes of finding them still alive and to encourage them to go down to the large bay, where I expected to find more, I gathered about a dozen of the shellfish. As the depth of the water passed my hips I thought I would try to get round by the rocks to near where the Captain and party had been left and then to return to them by going across the hills.

Climbing the rocks, then about twelve feet high, I soon found the weight of the shellfish and my wet clothes was likely to cause me trouble. I here found myself among some of the finest firewood I had seen up to that point, consisting of a stumpy growth of hard wood. So, having the matches with me, I lighted a fire. There is nothing more natural than that having got a fire I would like to try some of the shellfish. I pulled off my pants etc. and while they were drying roasted some of the shellfish. They proved to be so palatable that I am not quite ashamed to say that I couldn't resist temptation and they all vanished while my clothes were drying. They were at that time at their best, full of spawn and about the size of an egg and of nearly the same flavour. Having somewhat appeased my appetite and dried my clothes, thereby lightening them, I put a few shells in my pockets.

Knowing that I was about opposite where I had left the Captain and party, I started off with more enthusiasm than I had had up to that time. The hills were not very high at that point but the trees were of a scrubby nature and it was necessary to crawl under them many times. I didn't mind, as it didn't last long (the bush was not extensive) here there was no vegetation under them. However, I soon had to face an entirely different class of bush which was very thick for some distance. These thornless bushes were about four feet high and were so thick that it was only with difficulty that they could be broken through. Keeping to the direction I knew I had to go, I passed up the hill where I found them becoming lower and thicker until many times I got upon them and walked over the top of them, they being stiff enough to bear my weight. As I was rising the hill they become lower and ultimately none at all.

Reaching the top I had a good view of the country through which I had to traverse and found that it was not as far as anticipated and that I had taken quite the right course. There was a heap of stones or a cairn; (as the Captain had called them), quite visible on my right hand across one gully near the place where I'd left the Captain. From that time I had good travelling and went straight to the old camp.

It may be imagined what my feelings must have been then for it was with the gravest misgivings that I approached the place, knowing how I had left them. All the more so knowing all might have been dead by that time. It would be difficult to express either their feelings, or my own, at meeting again after twelve days absence. To my great surprise and pleasure I found them all alive, though some were in a bad state. It must be remembered that they had not had anything to eat but roots since they had a portion of the pig. This was positive proof of the substantial nature of their food value.

It is quite believable that some had scarcely moved since my departure. It is not necessary to say that they were pleased to see me and to hear the report of my discoveries since my departure or that the discussion lasted well into the night. After seeing the shells which I brought with me most of them were ready to take the advice I had to offer. I quite believe if I had asked them to start then they would have consented to go down to the big bay, though it was then nearly dark.

I explained to them what I proposed to do. The first thing in the morning I would take all who were able to travel down to the bay and see what it was like there. We would break down branches as we went as a guide for the return of one, whom I would send back to fetch the remainder as soon as possible. We were, as before stated, in a thick bush and did not know how far that continued. It certainly looked thick just there. We now numbered twelve on that hill. To wit: The captain, the chief mate, the second mate, the carpenter, the steward, Hipwell, Turner, Fritz, Harvey, myself and two boys, Liddle and Lansfield. It was finally decided that five would go with me – the Captain, the Mate Smith, the carpenter, Fritz, the Boy Lansfield and with myself as Pilot. They tried to get a little sleep but it was only a pretense of doing so as we had to keep the fire going with very inferior wood, but I had promised them better firing when we got down to the water.

The next morning we had not gone more than thirty or forty yards when we came into good open walking and were soon down to the water. Oh, if they had taken my first advice what a difference it might have made and more of them might have lived. They were now getting so weak that it was with difficulty that I could get them along.

We soon passed through some nice trees on the lower land with long, high ferns. These were always wet as a matter of course, but the high timber gave to me at least, some good encouragement. We found lots of dead-head wood as dry as the proverbial bone. This was a godsend to us after the terrible time we had had to keep fires going.

Where we had struck the bay may have been a quarter of a mile across with a rocky shore extending up to high water mark. The roots were plentiful all round by the shore, in fact on all the lower lands, as well as being of a better quality and growth. The rocks were sprinkled with limpets and it was pleasing to me how eager they were to gather them, and like me, they could not wait to cook them before having a feed.

While we were doing so, I saw a large bird about a hundred yards away sitting on a large rock about two feet above the water. Of course this looked too tempting to neglect. So, breaking a stick, I started after it as fast as I could until near it when I slowed and got to within a few feet. It sat there looking very simply at me until I let go the stick and it tumbled over. They almost screamed with delight.

We were not long in getting a fire and while I was plucking the bird they were roasting shellfish. Though it was not large it gave us a great deal of enthusiasm and it was a treat to see how they enjoyed their supper that night. So what with the limpets and the bird we were all in better spirits than usual.

It may seem strange when I say altho' it appeared to be a little more than a quarter of a mile from the bay to where we had left the others, that I did not send anyone back to fetch them. This is explained by that fact that travelling of any kind was exhausting and it was necessary to preserve our strength as much as possible as all were so weakened. Having plenty of protection from the wind and the weather and a comparatively dry place to lay down, there were few who would have cared leave such comfortable quarters after the way we had been living for the past three weeks. Self-preservation had become a vital matter for consideration. Though while still wishing to help the others, we spent the most comfortable night we had experienced on the island.

*   *   *   *

An excerpt from Raynal's diary, May 20, 1864:

The weather is variable, but generally cold and damp. The thermometer, on an average, has registered 3 degrees below zero at noonday in the shade, and at night is frequently below this.

In the shade! What a mockery! Are we not always in gloom and obscurity? The sun scarcely shows himself once or twice a week – just for a moment, between two clouds – and what a sun! so pale, so cold! And sometimes he does not make his appearance at all for fifteen days consecutively! Ah, but it is sad always to see above one's head an eternal veil of gray, a dull arc of sinister clouds! Oh, for more blue! for more sky!

One thing, moreover, has always produced upon me, as well as upon my companions, a still more painful impression, a kind of suffocating anxiety; namely, the monotonous and incessant beat of the waves upon the shore, at a few paces only from our hut, joined to the not less continuous murmur of the wind among the neighboring trees. It incessantly recalls to us our cruel destiny. Our nerves, therefore, are frequently over-excited; sometimes the most terrible melancholy, at once violent and gloomy, has been on the point of mastering us.[3]

The newspaper reports of the survivors' experiences range from the understated to the ludicrous. The source of these accounts seems to have been letters from the skipper to the company and a letter which Smith had written to his wife before returning home. Dalgarno says next to nothing about their experiences. The mate speaks of the six that went back to the wreck, but nowhere do either of them speak of the group breaking into independent foraging parties. This is an idea which developed at some stage and has been repeated through the years in various books and articles. Eden, in *Islands of Despair*, writes: "On reaching the harbour they ate shell-fish, and occasionally managed to kill a seal. It was at this time that the castaways separated into several parties, thinking that they would have a better chance of survival in this way."[4]

As we see from Holding's account, other than the group which went back to the wreck, all movements and separations were carefully planned, with the constant objective of searching for better living conditions. With each parting, there was an intention to reunite as soon as possible.

Some accounts had no more valid source than the imagination of the writer. The most flagrant was an article which appeared in the Christchurch *Chronicle* on 19 March 1891. The unnamed 'special reporter' on board the government vessel *Hinemoa*, during the search for another lost vessel (the *Kakanui*), filed a story mentioning the *Invercauld*:

We heard how the men dragged out a miserable existence day after day, then were seized with some strange madness which bred a strange resolve. They assembled one day near their rude hut and each determined to go his own way. They would make for the highest points of the island. Why? No one knows. There was nothing to be gained there, only hardship and exposure which no man could endure, so eighteen men went on their different ways and only the captain, the chief mate, and a boy remained in the hut, because they were too sick to move perhaps. Where the eighteen men went nobody knows.

In such a way are the legends of the sea created.

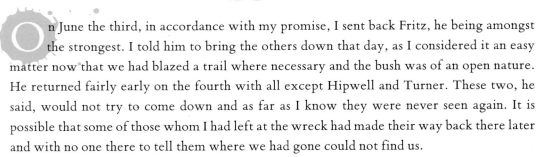

On June the third, in accordance with my promise, I sent back Fritz, he being amongst the strongest. I told him to bring the others down that day, as I considered it an easy matter now that we had blazed a trail where necessary and the bush was of an open nature. He returned fairly early on the fourth with all except Hipwell and Turner. These two, he said, would not try to come down and as far as I know they were never seen again. It is possible that some of those whom I had left at the wreck had made their way back there later and with no one there to tell them where we had gone could not find us.

We gathered some shellfish for them so they had nothing to do but feast for the rest of the day.

While Fritz was gone I hunted around to try to discover something new but nothing came of it. I did find that in going further down the bay it would be better to keep to the bush a little way back from the water owing to the amount of herbage there and the scrubby nature of the bush. The roots were very plentiful there, as well as the shellfish. While gathering some of the latter I saw a fish close to the rock on which I had killed the bird. The water being shallow and having heard that they could easily be killed by a light tap on the head, I got a stick and went gently into the water. Putting the stick near its head I just tapped it and had the pleasure of seeing it turn belly up. On looking round I found another with the same result. These might weigh half a pound each, so we were again in clover.

My fellow sufferers had now appeared to place more confidence in me and were prepared to take advice. Thus having now become the acknowledged leader I was not, or would not be, satisfied to stop there. I proposed to move again in the morning. After seeing the two seals

I was sure that we could get some more by doing so and to this they all agreed. As it was now getting night we prepared our sleeping quarters under logs with fire as near as possible and had quite a long chat deciding for the morrow. My proposition was to break camp early and with all hands, start for further down the bay. To do so I thought it best to go up the hill a little to the right of where we had come down, and cross over the hill on which they had been camped on so long. We would cross somewhat further down, thereby striking across a point to a bay that I had seen from the hills. So after spending another fairly good night we lost no time setting off on the morning of the 5th of June – we knew not where, nor what we were likely to meet with.

We could occasionally hear the pigs squealing but these did not frighten us; we would have been only too glad to see them. Though it may appear strange, it is a fact that we never saw another after the one the fellows had caught.

We soon found the travelling fairly easy and made good progress, soon finding ourselves near the top of the point. We had a good view of the bay and a fine sight of the entrance, which was perhaps three miles across with some islands on both sides. Although we could see that the bay widened out before us, we could not see the edge of the water nearest us on account of the bush. However we were soon to see it and to get an agreeable surprise. I noticed that some of the small brush had been cut and pointed these out to the Captain, (who was sticking to me). He had by that time recognized some of my abilities and now asked me how long I thought the wood had been cut. This was impossible to answer with any degree of accuracy, as the wood was unknown to me and climate has a great deal to do with the deterioration. This I explained to him but suggested that it had probably been cut more than three years, perhaps as much as five or ten.

This seemed to put fresh life into the whole of them, so much so, that all hands wanted to be first, especially those who had up to that time been so careless of their own interests. We followed the broken branches and could now distinctly see and trace, a track which we followed down the hill.

All eyes were now on the alert for signs of anything likely to be of service to us. We had only gone a few yards when the mate Smith (he being the tallest amongst us) called out that he could see a chimney near the water. It was soon visible to all. There was no restraining them or myself.

I said, "Where there is a chimney there must be a house"; and sure enough we could soon see a roof. It was but a short distance to it and all down hill, so there was quite a rush to be first there.

I have no recollection as to who was the first to enter, but the sight will remain on my memory to my dying day.

There was no door but a large doorway led into a good-sized hall or lobby. Beyond this we found the studding of a partition and a large fireplace under the chimney in what looked like a large kitchen. To the right of the entrance was the studding of another partition stripped of the boards, with a doorway passing into a large room. This might as well have been open.

On the left of the door there was a similar partition, also stripped of the boards and another doorway into another room. This was smaller than the one on the right.

As I shall later on have occasion to mention this timber, it will be as well to say the studding was all pine, size 2x4 while the boards were all hard wood. The roof was covered with shingles, all in a fair state of preservation.

There was really nothing very enticing about the house as a habitation but it would at least afford us the shelter of which we were in great need. All round the house there was growing a long flaggy grass and seeing some boards sticking up through it. I went to see what was there and soon made it out to be a kind of lean-to.

*   *   *   *

They had reached the Hardwicke Settlement.

[1]   Smith, A., *The Castaways*: p.15.
[2]   Bonnar, Lagos, Petersen, Corran, Genson.
[3]   Raynal, F.E., *Wrecked on a Reef: or, Twenty Months among the Auckland Isles*: p.196.
[4]   Eden, A.W., *Islands of Despair*: p.77.

13

Hardwicke

Sarah? Besson

" 'Cold, wet and windy' would be a succinct description of sub-Antarctic weather."

Sub-Antarctic Islands Guide Book

CAPTAIN Abraham Bristow, skipper of the Enderby whaling vessel *Ocean*, first sighted the islands on 18 August 1806 and named them Lord Auckland's Group after William Eden, the first Baron of Auckland, who was a friend of his father's. The next year he returned in the *Sarah* to begin sealing and named his anchorage in the vast harbour on the north-east side, 'Sarah's Bosom'. By 1823 the islands were to be found on Admiralty charts and by 1840 they had become an important stopping-off place for sub-Antarctic and Antarctic scientific expeditions.

In this one year, ships of three Antarctic expeditions landed on the Aucklands: Wilkes in the *Porpoise*, D'Urville with *L'Astrolabe* and *La Zelee*, the latter leaving the same day that the British Antarctic expedition under Sir James Clarke Ross, en route to the Antarctic, made landfall. His ships, the *Erebus* and *Terror* gave their names to the two small coves on the east side of Sarah's Bosom, which he renamed Ross Harbour.

As a Canadian, I could not help but note the fact that these two coves bore the same names as the ships of the Franklin expedition which were sent to search for the Northwest Passage in the Arctic. The connection was interesting. Sir John Franklin had been Lieutenant-Governor of Van Diemen's Land (now Tasmania) from 1838 to 1843 and in this position had shown considerable interest in the Antarctic expeditions. When Ross returned, he wrote to Franklin suggesting that the Aucklands would be an excellent location for a penal colony. As transportation to New South Wales had recently been abolished, convicts from all over the British Empire were being sent to Tasmania and Ross suspected that Franklin might be interested in the development of another location. Nothing came of this project, but the idea caught the imagination of Charles Enderby, of the major British whaling firm Samuel Enderby & Sons, who shortly afterwards secured the islands from the crown as a rendezvous for their South Sea fleets.

A local paper reported Franklin's return to England: "...thus departed from among us as true and upright a governor as ever the destinies of a British colony were entrusted to."[1] Two years later he was offered the command of an expedition to search for the Northwest Passage. Franklin set out with 134 officers and men in two ships, the *Erebus* and the *Terror*. It is unclear as to whether these are the same ships which had been under the command of Ross in the Aucklands, or simply named after his ships in recognition of the Antarctic voyage. A comment in an 1898 edition of the *Encyclopaedia Britannica* implies that they may have been the same ships. "...Sir James Clark Ross, – the latter commanding the *Erebus* and *Terror*, with which the name of Franklin was to be for ever so pathetically connected."[2] James Clark Ross was later sent to the Lancaster Sound area in the search for the ill-fated expedition.

Not only was Charles Enderby appointed Resident Commissioner of the Auckland Islands, but Lieutenant-Governor as well, which in essence gave the Aucklands the status of an independent colony. Enderby had read Benjamin Morrell's 1829 report: "Auckland's Island is one of the finest places for settlement. There are very few spots that could not be converted to excellent pasturage or tillage land."[3]

In his prospectus, Enderby did say:

Although we have no reason to doubt the accuracy of Captain Morell's statements, yet, to guard against the possibility of their being overrated, and thus lead to disappointment, it might perhaps be as well to receive some of them with caution. Our object is not to mislead, but to state such facts as we are able to collect from our own officers and others who have visited the islands. Their accounts of them, however are substantially in accordance with those of Morrell.

It is true the Captain was at a different part of the island from the other authorities hereafter quoted, and also that his visit was at the height of summer, and that he only remained 8 days; whilst Sir James Ross remained there 22 days at a period corresponding to our May, and Admiral D'Urville and Commander Wilkes were there, the former nine and the latter three days, at a period corresponding to our October.[4]

Despite his best intentions, the reports did mislead – terribly. Based on the glowing reports, he recruited settlers to fulfill his dream "to colonize and establish a prosperous settlement, to act as a whaling base, refit and replenish trading and whaling vessels and trade with the separate colonies of Australia and New Zealand."[5] A less appealing area for farming could hardly be imagined. Full of hope, and attracted by Enderby's glowing prospectus promising fertile soil, healthy climate and "freedom from aborigines", the settlers were in for a very nasty shock.

One hundred and six colonists including "medical men, clerks, a surveyor, a storekeeper, boatmen, coopers, shipwrights, carpenters, smiths, brick layers, masons, agriculturists and labourers, besides sixteen women and fourteen children" were aboard the *Samuel Enderby*, *Fancy* and *Brisk* when they set sail from Plymouth on 18 August 1849. The *Samuel Enderby* was the first to arrive 118 days later, sighting the islands on 2 December 1849 and entering Port Ross two days later. The *Brisk* arrived on the 11th but the *Fancy* did not reach her destination until 27 December after almost missing the islands in the fog. Sarah Cripps had the misfortune of being aboard with her husband and her three children aged four, three and one year. "For the whole long voyage she had been so desperately sea-sick that at one stage she begged the sailors to throw her overboard."[6]

SAMUEL ENDERBY.

Rather than arriving at an uninhabited island as promised, the settlers (and Charles Enderby) were greeted by a boat manned by three or four Maori naked except for a piece of sealskin wrapped around their waists.

Mr Thomas Younger (whose infant daughter was to die at the settlement of convulsions) described the scene:

> They were painted and had feathers in their hair, and had one woman amongst them. One of them came on board whilst we were still outside. We had a Maori sailor on board, who we asked to pilot us in, and from that day he was called Pilot Jack – a savage looking Maori he was. The sailors first took him and dressed him in some of their clothes; he then took charge of the ship and piloted us right in. We anchored between Shoe Island and the (Maori) settlement.[7]

Accounts of these Maori (with their Moriori slaves) give the number greeting the *Enderby* settlers as 47: 20 men, 17 women and 10 children. They are described as being of a "fugitive band who made no claim to the land".[8]

These numbers are in contrast to a report found in the *Journal of the Polynesian Society* (1893) which states that there were "seventy of these people on the islands, who had been brought by a Colonial vessel from the Chatham Islands, about eight years previously; thirty of them under a chief named Matioro on Enderby Island; twenty-five under another chief named Manature and the others were independent."[9]

The arrival of these Maori from the Chatham Islands in 1842 and their departure shortly after the closure of the Enderby Settlement is another tale.

The site was surveyed and building locations established. Since many of the buildings had been partly assembled in England, basic living quarters were soon erected and on New Year's Day, amidst celebrations, the settlement was christened 'Hardwicke', after the Governor of the company. Work continued and by late summer 18 buildings had been completed, including Government House (a large two-storey building), the barracks for single men, a chapel, a zinc-roofed public store, workshops, sheds and houses. A jail was erected on Shoe Island.[10] Built with Maori help, half a mile of road stretched along the shoreline. A battery of four guns defended the settlement.

Departure of the whaling ship 'Samuel Enderby' for the Auckland Islands.

Three horses (for which they never found a use), 45 horned cattle, 258 sheep and 51 lambs, "in addition to some hundreds of pigs (which were allowed to run wild over the islands)"[11] were shipped in from New Zealand.

When they had first arrived, the Maori had a number of hogs,[12] and had "enclosed and cultivated a considerable quantity of land". Enderby, to avoid disputes, used his powers and "deemed it advisable to compensate them for what they considered to be their rights, and they surrendered to him all claims to any hogs and to all the land, whether enclosed or otherwise".[13] He also hired all the Maori "in the Company's service, the two chiefs as constables, the other men as labourers, at moderate rates of wages, but quite equal to agricultural wages in England."[14]

The two chiefs were instrumental in promoting good relations between the whites and Maori,[15] but difficulties arose from time to time. Thomas Younger described the night that "so many [Maori] came that we were under arms all night, armed with flint-lock Brown-Besses – all passed off however without disturbance – there were fully thirty or forty men that came over at that time, but they had no guns with them..."[16]

Charles Enderby was driven by a dream and one can't help but wonder if he was in another part of the world from his "citizens", for the reports sent back to England bore no resemblance to the harsh reality the settlers were experiencing. He raved about potatoes and cabbages as being "of a quality not to be surpassed", and of the soil being "rich beyond comparison".[17]

Edmond Malone, visiting the islands on the HMS *Fantome*, described a depressing scene:

The farms, and there are several, are everywhere failures; nothing grows to any size, although every care seems to have been taken. The potato and vegetable gardens are fenced round with stakes; and every 14 feet inside with the same, to keep the wind off, looking like sheep-pens, but all to no purpose. The potatoes are about an inch and a half in diameter, and bad; and the turnips run down like miserable radishes: the only vegetables we ever got there were, once or twice, some of the leaves, dressed like spinach. The Maories, at a small pah of theirs at Ocean Point, grew the best cabbages and turnips, but they were good for nothing.[18]

The weather was appalling with little sun during the summer and constant gales with rain and fog throughout the year. It soon became obvious that the project was an unmitigated disaster. Not only were the settlers disappointed by the realities of the climate and soil, but the promises of rich sealing and whaling also came to naught. The seals had been hunted almost to extinction in the 1820s and no whales were ever captured in the bays where they were expected to come to calve.

Five of the married couples left in May of the first year and two special commissioners from England, representing the shareholders, arrived shortly afterwards to investigate the lack of returns. As a result of their inquiry, HMS *Fantome* arrived in 1852 to arrange the winding up of the settlement. On 5 August, hopes dashed, the settlers departed – most to return to England or Australia with only one or two settling in New Zealand. The prefabricated buildings were taken to New Zealand where some remain in use to this day. The assistant commissioner William Mackworth wrote in his diary the day he left: "The satisfaction I felt at this moment is beyond description. My miserable life at Port Ross will never be forgotten".[19]

* * * *

As I stood beneath rata branches in a Tolkien-like forest, it was almost inconceivable that 150 years before, a settlement of over 100 people had existed in this area. There was a large two-storey house for the Governor, a zinc-roofed public store, chapel, workshop, cottages for the married couples, in all about 18 buildings. All that remains are vague indentations in the earth. It is just possible to see the route of the old road along the coast where a few pebbles lie partially covered by the deep peat.

The last permanent building, possibly the single men's quarters and undoubtedly the one in which the *Invercauld* survivors took refuge, was demolished by the coastwatchers in 1941-42 to make their shelter in Ranui Cove south of Port Ross. Unfortunately, I had not realized that that building had been standing so recently, so we missed the one location at the old Hardwicke site which could have been identified as having been directly connected with the survivors.

Holding mentioned seeing a long spar standing in a clearing, which must have been the original flagpole of the settlement. Although he makes no further comment, Smith described how they cut it up to make a raft. Three years after the rescue of the *Invercauld* survivors, the *Amherst* put in to Port Ross.

Armstrong writes:

The day following we formed the first depot on Enderby's Island in the hut nearest the beach (occupied by Teer and four others) ...The case No. 1, containing clothing, blankets, compass, matches, tools, &c., was placed in a good position, and on it I wrote – "The curse of the widow and fatherless light upon the man who breaks open this box, whilst he has a ship at his back." A spade was left with the case. A bottle suspended over it contains a letter of which I give a copy, it being similar in its contents to those left at the other depots.

[In this letter he says:] "There has been left here today, by order of the Government of Southland, a case (hermetically sealed) containing absolute necessaries for the use of castaways. I need not add exclusively for their use, for surely no one with a ship at his back will have so little respect for his manhood as to take aught of what is contained in this box. Three similar depots will be made in other parts of the island. One at Port Ross, one at the head of Saddle Hill Inlet (third bay south from this,)[20] and another in Carnley Harbor, (the Straits.) Those who may come after, I will ask, in the name of suffering humanity, to see that the cases are preserved from injury, and that the landmarks remain firm in their places. Should the lettering on the boards be indistinct, pray renew with paint."[21]

They marked their visit by carving the details onto a triangular board which they nailed to a new pole.

We erected a signal mast, a fine spar, perfectly sound, forty-five feet long, one foot in diameter, which we found on a beach, having apparently formed a portion of the old jetty. It is placed in the position formerly occupied by Mr. Enderby's flagstaff, stands thirty-eight feet above the ground, and is properly set up with shrouds. On its top is a small barrel; below it appears three boards, each eleven feet long (lettered) in the shape of an equilateral triangle. The whole painted white presents a very conspicuous and sightly object, which can be seen from all the neighbouring hills, and from the entrance of the harbour. At its foot is a small, weather-boarded shed, closed in, containing the case, a spade, and a small box of books. Fixed to the mast by a band of copper band is a bottle, holding a letter."[22]

The 'Amherst spar' now leans precariously against the highest branches of the surrounding forest. A pig's skull, placed on a portion which has broken off the top, seemed to say it all – eradication and decay.[23]

Not far from the spar an old tree stump about six feet (2m) high huddles under a corrugated iron A-frame roof. Sailors from the Victoria had carved their 'Kilroy was here' graffiti:

H.M.C.S. VICTORIA, Norman,
In Search of Shipwrecked People
Oct. 13th, 1865.

along with the names of several of the crew. No crude scratches these, but carved with care and with a flourish.

Great strides have been made in the preservation of artifacts but unfortunately serious errors have been made. The stump has deteriorated since the protecting trees around it were cut, allowing sunlight and moisture to wreak havoc upon the ancient wood and the later installation of the protective roof has only exacerbated the situation by allowing the wood to dry out. The stump has now been treated with a wood preservative but the damage has been done. The writing is fading and soon, this too will be gone. These two artifacts bring into focus the question of preservation versus natural decay. Some items from the islands have been brought to the mainland (for example the figurehead of the *Derry Castle* and the coracle built by the *Dundonald* survivors), but the principle of leaving items in situ is now the norm and within another generation only written records and photographs will remain of this southernmost colony of the British Empire.

The policy of the Department of Conservation is to use every means at their disposal to hasten the recovery of the islands to something approaching their original condition. Through careful management and the eradication of introduced species, the vegetation is beginning to return to its natural state. The philosophical question is: when is something considered intrusive? A case in point is the *olearia*, or tree daisy, which occurs naturally on the Snares and on Stewart Island, but is not considered native to the Aucklands. Whether it was brought on the foot of an albatross, or introduced by humans in the 19th century makes all the difference in how the policy of control is applied. The current thinking is that it was introduced by man, thus the concern about its spreading has resulted in a policy of limited control in the Port Ross area and eradication elsewhere.

Rules for the traveller in the Aucklands are very strict. So as not to inadvertently transfer seeds from one island to another, all boots must be thoroughly scrubbed with a strong wire brush dipped in a chlorine solution. Pockets, cuffs and folds in gaiters must be carefully checked for any seeds which may have been picked up on a previous trip.

We made our way along the shoreline through the site of the Hardwicke settlement. The only remaining building is the old depot shed filled with the detritus of ages: an old spar, a fishing net, two old bottles, a rusty adze head, items found and pitched there by various visitors. In 1982 an archaeological survey was done and a detailed report made available. Unfortunately, this report drew the attention of collectors to the islands and many unauthorized visits were made, resulting in a virtual cleaning out of the site.

To take in the whole situation, I will now try to describe the place as we saw it. We were in a corner of a bay, with the shoreline on the left running straight down the Bay. This bay appeared to extend seaward for six or seven miles without obstruction of any kind and looked to be three or more miles wide at its mouth; from what we had already seen from the hills there were islands on both sides. Extending down to the water, was a good sized clearance of about four or five acres. The timber had been cut and now all covered with the long grass. At the back of the house there was the trail which we had come down, by the side of which some told me after, they had seen some graves. There was a small creek in the left hand corner with a small bridge across it.

* * * *

The landscape was exactly as Holding had described it, although of course, the rata forest has claimed the land which was then grassland. To protect the fragile soil, the trail to the tiny cemetery has been recently covered with duckboard. The cemetery is, as cemeteries tend to be, a sad place. As we stood, pondering the eight lonely graves, it was not hard to imagine the mournful processions making their way up the wild hillside. The silence, broken only by the wind and the clear, haunting trill of the bellbird, wrapped itself around us. Early photographs show leaning headstones in a ravaged area, but today the graves have been restored and a chest-high wire fence guards against the inroads of vegetation, pigs and sealions.

Charles Eyre, a survivor of the *Dundonald* wreck, had stood here with his companions in 1907 when they first arrived at the Hardwicke site. Later, they were to bring the body of their first mate, Jabez Peters, from Disappointment Island to be interred. His grave is clearly marked.

I think the most touching and solemn discovery which we made during our whole stay on the island was that of a little cemetery not far from the depot, in which were sleeping the bodies of unfortunate mariners who, from time to time, had been cast ashore there. It was as sad and solemn thing to stand there in that silent and deserted spot, and contemplate the last resting place of those for whom friends and relatives had waited and mourned in far-off lands."[24]

We had come to see one particular stone:

SACRED
TO THE MEMORY OF
JOHN MAHONY
MASTER MARINER
SECOND MATE OF THE
SHIP INVERAULD[25]
WRECKED ON THIS ISLAND
16 MAY[26] 1864
DIED FROM STARVATION.

In 1889, for the sum of £3, a stonemason by the name of John Fraser had been commissioned by the Invercargill Wreck Fund to carve a marble tombstone for Mahoney's grave. This was to replace the original board inscribed UNKNOWN which had been hung in a tree by the sailors of the *Flying Scud* who had discovered his body in 1865. Mahoney has to be the most remembered man, for at one time there were three, if not four markers for his grave.

A replica of the memorial placed by the *Southland* is now fixed to the fence behind the stone. Eyre describes the original:

Over the next grave to this there was a tombstone constructed of rough pieces of wood nailed together, and on it there was painted in black letters, which was evidently the work of some sailor, the following inscription, which I reproduce almost exactly as it was on the board:

"Erected by the Crew of the S.S. Southland over the remains of a man who had apparently died from starvation and was buried by the crew of the Flying Scud. 3rd Sept., 1865."[27]

In fact, the *Southland* party exhumed Mahoney's corpse from where it had been buried by the *Flying Scud's* crew. The captain had a coffin constructed and Mahoney was reburied in the cemetery.

The only original headstone still legible from the time of Enderby Settlement is that of Isabel Younger, aged 3 months, who died 22 November 1850. Her father was the civil engineer of the settlement and unwilling to have his child forgotten because of the decay of a wooden marker, took a grindstone and carved the inscription around the square hole. The other known graves are those of Janet Stove, aged 14 weeks who died 10 October 1851, and John Edward Downs, 21 February 1852. It is thought that David McClelland, aged 60, a *General Grant* survivor who died 23 September 1867, was also interred here. A seaman by the name of Thomas Cook is remembered, but it is not known whether he was actually buried in the cemetery or whether he died at sea. If this is the case then there is one grave unaccounted for. The exact location of some of the graves is in doubt due to the deterioration of the markers over the years.

R.A. Falla, in *Comments on the Enderby Settlement and the Cemetery at Port Ross*, recommended that a permanent Historic Places plaque with the names of all the burials should be erected nearby. The graves are now all marked, but there is no indication of any further work having been done.

The Department of Conservation states clearly in its management plan that the "primary source of awareness should be focused on the off-site awareness" and that on-site interpretation should be limited to material provided to permitted visitors and "no on-site interpretation is appropriate". However, the graveyard is a present historical site and a memorial to the 124 souls who died on the island or in the wrecks would not be inappropriate.

Wanting to see more, the second mate and I started off along the shore. We hadn't gone more than a quarter of a mile when we observed a long spar which had evidently been intended for the Mast or Yard of a vessel. The shore here was a sand beach for a short distance then became shingly. If memory serves me right there was a small island about the centre and a few yards out.[28] The shore then became rocky to the point. There was directly at the point some ragged looking rocks projecting out into the water for perhaps twenty yards. These were very irregular with a few boulders scattered over them. On walking out on the rocks, I saw at a distance a bird like the one I had previously killed, so I had to try to get that. The only stick I could get was a rotten one, but quite good enough for the bird. But before I got near the bird, Mahoney called out, "Bob, for God's sake, come back, here is a Seal!" So I, of course, lost no time in getting back to him, but could see no Seal. I said, "Where is it?" and pointing to the water beside the rocks, he said, "There, in the water".

Sure enough, there was a seal not more than two feet from the rocks laying quite calmly in about four feet of water just poking up his nose occasionally to breathe. As I have before stated, I had had no previous experience with them and neither had Jim. The rotten stick was the only implement at hand but it had to be made immediate use of for fear of losing him. I knew enough to be sure a blow under water would be the loss of him. So taking advantage when he rose I brought down the stick on his nose with all my might. I was horrified to see and feel the stick break.

I called to Jim, "For God's sake, give me some Stones", and grabbing some myself began to pound him on the head. This kind of treatment was too much for the Seal. We could just reach him and taking hold of his flippers we lugged him onto the rocks. We were none too quick as he soon appeared to be recovering, so we turned him on his back in a crevice from which he could not escape. When we got him on his back I took my knife from the sheath and tried to plunge it into its breast. The breast and neck of the seal, or the sealion we must call them, carries the thickest part of their skin. My knife had by this time become rather dull on the point and as a consequence instead of going through stopped and my hand slipped down the blade cutting two fingers to the bone. (These marks are plainly visible today though they are nearly fifty-five years old.) I then drew the knife along his throat and pushed it home. The seal closed up the wound by squeezing himself together but he could not stay the bleeding inwardly.

I happened to have a handkerchief, so bound up the fingers and started for the house telling Mahoney to wait there and I would send someone to help him to get some liver and fat. It's needless to say there was soon some help there but as it was now getting dark they brought only what I told them to bring.

Some of the men had been clearing out some of the rubbish and in doing so had found a piece of galvanized sheet iron about eighteen inches at one end, tapering to about three inches at the other. We were able to use this as a frying pan. There was no stint as to what anyone could eat on that occasion and I think I can safely say that all enjoyed their supper. We certainly had good cause to be thankful for the advantages obtained on the fifth of June.

While we were looking for something to eat some had been busy preparing their sleeping quarters and were ready to lay down when they had their fill. Three of us chose to occupy the lean-to.[29] We cleared off the long grass and found that the front had fallen in, (or perhaps had been knocked in), but this would act as a floor. It now being quite dark, we just crawled in as best we could. We were, however, more comfortable than at any previous time since we were wrecked and I think I can safely say we all slept better. The only persons who had anything to cover themselves with were the Captain and First Mate for when escaping from the wreck they had thrown off their Seaboots but had stuck to their overcoats.

June the Sixth found us early astir, but many of us had a very severe headache. I suspected the cause, for I had noticed who had eaten of the various parts of the sealion at once came to the conclusion that it was the liver which caused it. So we ate no more of that and we had no more headaches after that passed off. We went after the remains of the seal, skinned it and brought it to the camp.

After a good breakfast we began to explore the neighbourhood for anything that might become useful in future. We searched every nook and corner for anything we could find, and I think I can here relate to a nicety about what was found in toto. Having mentioned the piece of galvanized iron, I may now include another piece of similar size. There was also one half gallon water can with bottom rusted out into which I afterwards put a wooden bottom. There was one old axe of English make with the eye bursted [sic], one old adze with the round eye and half the face gone and two or three pieces of fencing wire up to two feet long. There were several preserved meat tins of four and six pounds capacity, two or three pipes and one old spade, two bits of roof slates and five or six bricks. I think this about covers our discoveries there.

While we were busy searching amongst the long grass we discovered a pebble walk. This,

I should judge, to be about four feet wide, perhaps eighty yards long and very nicely laid. There were also many small patches which looked as if they might have been laid out for gardens. They were of a great variety of sizes and had been fenced in by laying one bush on top of the other. These fences were quite visible. We now had every proof that the Island had been occupied for a considerable time. This, of course, gave us great hopes of their early return. As a lot of information in reference to these islands (in which I had so much interest) has since come into my hands, I hope to be able to explain a good deal that was to us nothing but a mystery at that time.

The water was shallow for some distance along what we termed the Beach, thus there was not much prospect of getting shellfish. We did not mind as we had the seal meat and a plentiful supply of roots. It would perhaps be as well for me now to give a better idea of what these roots were like. They varied in size and utility, according to the locality of their growth; on the hills they were short and tough while those grown on the lower ground were longer and more tender. The longest grew to about eighteen to twenty inches; they were of a scaly nature to the eye, and every year's growth formed a knotty fibre. The petals or stems were about $^{3}/_{4}"$ in diameter. The leaf was similar to the marshmallow, but over a foot across. In the seeding time they threw a cluster of green betties which turned black on ripening. When peeled, the flavour somewhat resembled that of the sweet turnip, but as it was seldom they grew more than two inches in a season, they were pretty tough. And very binding on the bowels."[30]

On the second day we were at the house, a fine Tabby tom cat made its appearance, and by degrees got sufficiently tame to come quite close to us. But the second mate threw a brick at it and we never saw it after. He wanted to kill it to eat.[31]

I might here be allowed to state that in consequence of our being compelled to lie so close together we contracted vermin. As we got oil from the seal I rubbed some on my head; this I found effectual. I pulled off my clothes one article at a time and scaled them, thereby getting thoroughly clear of them.

As some had been barefoot except for rags, the seal skin was used by any who liked to make a kind of moccasin which soon disposed of the hide as the seal might of weighed 120 lbs. This may in a measure account for their not having tried to get about more. About the eighth of June the carpenter and two   others, who were strolling round the shore, killed another of about the same size but of quite a different kind. It was I believe, what is called the fur seal which was of a lighter colour and spotted.

I discovered later that the captain had been keeping a kind of diary on the margin of a Melbourne newspaper, which he happened to have in his pocket when we were wrecked. I had no knowledge of that at that, or any other time while we were there.

It would become monotonous for me to attempt to wade through our daily occurrences, as there was so much similarity. About this time I decided it was to our advantage to go further seaward. I placed the matter before the others and asked for volunteers, but could not get one. My mind was made up and on the following morning I started out after telling them that I would go as far as I could and let them know the result on my return, if I lived to return.

1   *Encyclopaedia Britannica*, 9th ed. Vol. XXI. (1898): p.720.
2   ibid: p.720.
3   Department of Conservation, *New Zealand's Subantarctic Islands*: p.15.
4   New Zealand Pamphlets, Vol.2: *The Auckland Islands:A Short account of their Climate, Soil and Productions; and The advantages of establishing there A Settlement at Port Ross for Carrying on the Southern Whale Fisheries*. Charles Enderby, Esq, F.R.S. London.
5   Pope L., 'Auckland Islands – Wild Splendour.' *New Zealand Geographic*. Oct.-Dec., 1990: p.84.
6   Macgregor, Miriam, *Petticoat Pioneers: North Island Women of the Colonial Era*, AH & AW Reed, Wellington, 1973.
7   *Journal of the Polynesian Society*, 1893, p.84.
8   Department of Conservation, *New Zealand's Subantarctic Islands*: p.26.
9   *Journal of the Polynesian Society*, p.84.
10 The only occupant was Mr. J.S. Rodd, the chief surgeon, who was confined several days for drunkenness.
11 McLaren, Fergus, *The Eventful Story of the Auckland Islands*, p.56-57.
12 They had brought 150 hogs with them.
13 *Journal of the Polynesian Society*, p.84.
14 ibid: p.84.
15 Thomas Goodger named his son, born 22 Oct. 1850, after chief Matioro. (He later changed his name to Mathew.)
16 ibid: p.85.
17 Department of Conservation, *New Zealand's Subantarctic Islands*, p.16.
18 Malone, Edmond, *Three Years' Cruise in the Australasian Colonies*, p.64.
19 Pope L., 'Auckland Islands – Wild Splendour.' *New Zealand Geographic*, Oct.-Dec., 1990: p.86.
20 Norman's Inlet.
21 Armstrong, H., *Government Gazette, Province Of Southland*, Vol. 6, # 9. 11 April 1868, p.53.
22 From Armstrong's report.
23 When we returned in 1995 another section had broken off the top. The triangular sign (obviously a replacement of the one described by Armstrong) has been removed to New Zealand for preservation.
24 Escott-Inman, H., *The Castaways of Disappointment Island*, p. 280.
25 Invercauld is spelt incorrectly on the stone.
26 Mahoney is incorrectly identified as John, and the date is in error.
27 Escott-Inman, H., *The Castaways of Disappointment Island*, p. 281.
28 Davis Islet.
29 From Smith's account this would appear to have been Mahoney and one other (Holding).
30 These roots in all likelihood were *stilboacarpa polaris*. They are very rich in vitamin C, which would account for the survivors not suffering from scurvy.
31 There are still feral cats on the island.

14

seals

distant island

I crossed the small creek on the foot bridge and soon found a trail similar to the one we had followed to the house. I knew it must lead to somewhere so this I followed, caring not where it went so long as it was possible to get something to keep body and soul together. I found the trail easily followed and in about a third of a mile came out into a clearing at the head of a very small bay with a sand beach in front. To my surprise there stood the skeleton of another house standing on to the water and perhaps fifty yards therefrom. The framework did not appear to ever have been finished as there had been no boarding work done to it but the frame might have been covered at sometime with a tent or canvas.

There had been a garden round by the building and potato tops were still plainly visible. The potatoes were, of course, not very large – in fact only the size of marbles. There was also some cabbage which had gone to seed. The seed was not ripe and the cabbage had never been of use, having run wild.

Though all this was useless for the time it was certainly of an encouraging nature. I set off again following the trail and I don't think it was more than half an hour before I came out on open land. I continued to cross it towards the shore and along the edge of the rocky bluff I saw plots like those I have before mentioned. These were fenced just like the ones at the camp and I am as much in the dark as to their purpose today as I was then.

The bluff here was about two hundred feet high [Dea's Head] and I now had a good view of the sea looking to the Northwest and over the lower reaches of the bay. The point of land beyond the small bay cut off my view towards where we were wrecked. The land near the water's edge, about seven feet above the high water line, was fairly dry and level. Having seen all there was to be seen there I turned my steps seaward again, steering for the further point and was soon in low bush. As before stated it was bad travelling but fortunately it only extended a short distance until I entered the higher timber.

It may be of benefit to someone if I give a description of this timber as the best of my ability will allow. The hard wood is of a very scrubby nature with the roots above the ground and badly tangled so that it was often necessary to go along on hands and knees. The roots branch in all directions above ground and are much entwined. The colour is of a reddish tinge and the wood very hard and solid. I have often thought there was a fortune for any one with sufficient capital to fetch a shipload as an experiment for use in all sorts of furniture and fancy goods. I would be negligent if I did not here describe the foliage which was very thick, so much so that it was with difficulty that the sun was seen through it. The leaf was about the shape of the bay leaf and a beautiful green. In the spring of the year it bore a pink flower resembling the flower of the Fucher [sic] turned upwards. Altogether a sight to behold.

(I would like to say that I have in my possession a book,[1] the author of which represents as having gone over this very ground, and that there is a wide difference in the time he states it occupied them and my time. I think probably there is quite as much difference in the impression we each wish to convey.)

To go back to my trip; to obtain better travelling I had taken a course a little to the right of where I had intended. This brought me into the corner of a bay with a rocky shore all round. Here I got the best view of the bay that I had seen up to that time. I followed the left hand shore for a short distance but finding the rocks very slippery took to the bush which soon became very thick and scrubby; so thick in fact, that I could not get through it. Finding that it was easier to crawl under I took that course. As I was crawling along I had only gone a few yards when I was startled by a seal springing up in front of me. I don't think I was frightened, but certainly surprised. Although it was impossible to catch it in that bush it was satisfying to have found one of their haunts and it give me great hopes for the future. This was what I was looking for and another would not surprise me in the least.

This bush did not extend so far and I was soon in more open country – out on a place bare of bush and timber but covered with a long flaggy grass. There was nothing to indicate that anyone had ever lived there but to my surprise and pleasure found any amount of traces of seal. It was now time to look for a place to lay down for the night and as my clothes were all wet I soon made a fire. I could get water at any time for it lay in ditches everywhere. Firewood was plentiful, the trees being low and of a scrubby nature and I could break off what I required and it was little trouble to keep the fire going for the wood was very hard. This was a good thing as the matches were by then getting very scarce. The soil here was of a peaty nature (as in fact it was all over the island) but here, being six or eight feet above the water, was comparatively dry. So after getting some roots and some limpets, which were plentiful all around the rocks, I laid down for the night. There was little to be gained by trying to choose a place to lay as they were all about on a par with everything covered with decayed leaves. It must not be supposed, however, that it was all sleep, for the fire had to be frequently replenished, but having become quite accustomed to that by this time, it did not worry me much. It may be easily imagined that we had always been subjected to sleeplessness owing to our thoughts being far away, and consequently, subject to dreams of everything and everybody we had known.

* * * *

During the early 1800s the only indigenous mammals (whales, sealions and fur seals) were hunted almost to extinction. The Hooker's sealion, the world population of which centers on the Auckland Islands, now numbers about 5000 which have been fully protected since 1881. November marks the beginning of the breeding season with the arrival of the bulls to stake out their territory. By the time we arrived on 3 December 1993 each cove had its "beach master" – an adult bull big enough and experienced enough to claim it for his own. Sealions are usually associated with the shoreline area, but we were soon to learn that they spend a good deal of their time well inland.

Holding mentioned coming face to face with a sealion and although we didn't have that 'pleasure', we did follow their low, narrow trails which led from the beach into the woods. We were amused to learn that sealions do not like the feel of falling rain on their bodies and will seek shelter in the bush. Our first inland sealion came as a great surprise. Leaving Hardwicke behind, we crossed the stream which Holding mentioned. Those of us with leather hiking boots did so by means of a moss-covered log; those with rubber boots waded boldly across. We had just come up the gully and there, lying in splendour, was a beautiful specimen, weighing probably 500 kilos. He was big, black and sleek, and not sure of us at all. Standing well back, we observed him as he observed us. Paddy dropped behind a log about 20 feet (6m) off and began shooting pictures. Kodak stocks must have had a sharp upswing after we returned to New Zealand, for I am sure that between all of us there must be two or three hundred pictures of that one seal, and many feet of video film. My film takes several sudden swings as I quickly retreated from his periodic feints. He would roar, advance several feet, pause and go in for several minutes of head swinging, then repeat the behaviour. At one point he lay down like a gigantic dog and rubbed his head on the moss. We relaxed, he didn't, suddenly rising up on his flippers and charging towards us. Paddy was the bravest of us, holding his ground and clicking his shutter like a motor-drive. My husband, knowing there would be plenty of shots of the sealions, came away with a wonderful collection of pictures of the photographers.

Eventually deciding that we were not a threat, he turned and lumbered back towards the sea, about 150 feet (45m) away, pausing en route to have a rest while we waited to see what he would do next. Then slowly, ponderously, he slipped down the slope onto the rough shingle beach and into the sea. We stood and talked about the experience and about sealions in general and before we knew it, he reappeared twenty yards (18m) further along to have another look at us. We accused Pete of not paying him for his performance. Pete explained that the young bulls, who did not have their own harem, tended to be much more aggressive than the old bulls who had no need to prove themselves. They are normally nine or ten years of age before  they are capable of holding on to their own territory, so there are a lot of young bulls trying to establish themselves. This was a young one who was feeling 'really cocky' at having held so many strange two-legged creatures at bay, and he was just coming back to check that we were indeed on our way. Not having a kipper to flip to him we left him to his hopes of holding this beach and proceeded on our way.

Pete reminded us again of the importance of not being so taken by what was happening in front of us that we forgot to keep an eye on who, or what was behind us. A warning which we were well to keep in mind over the next few days. In the next bay we met an even more aggressive chap who was simply not going to let us pass. Here in Terror Cove the bush was particularly dense and we did not want to have to bushwack to get along the coast. Some of our group were more courageous than others and stood their ground when charged by this determined fellow. I stood well back, making good use of the zoom feature on my camera. We retreated in two parties under the watchful eye of Pete – and the sealion. Unfortunately, the animal cut short our appreciation of the German Transit of Venus Expedition site.

This 1874 expedition was important, for from measurements taken from various sites in the southern hemisphere it would be possible to work out the distance of the earth from the sun. The opportunities for this were rare for the next chance would be in 1882 and the following not until 2004. All that remains is the brick plinth on which they placed their apparatus.

Cutting up through the woods, we saw a patch of ground that had been well 'snuffled' up by some of the wild pigs which still remain on the island. Pigs have been responsible for the greatest modification to the ecology of the main island and in accordance with the Management Plan attempts are being made to exterminate them. "Extermination will not be easy and will require substantial resources and commitment given the large size of Auckland Island, its inaccessibility, rugged terrain, and in parts impenetrable vegetation cover. Research will be required to develop and test a method of extermination that is effective against pigs but not a hazard to non-target species."[2]

Down on the next beach we found another even more belligerent youngster whom Pete had to ward off with a stick as we passed by to Dea's Head, a pinnacle of basaltic columns which rises 200 feet (60m) from the sea. We scrabbled up what seemed an almost vertical slope which was covered with dense vegetation. Breathless, on hands and knees for the last 30 feet (9m), I became acutely aware of being watched. Turning my head, I found myself staring straight into the milky eyes of a young female sealion, not more than two paces away. She had stayed quiet as the younger members of our party had passed her. She lay quietly as we took yet more photographs. The magnificent view behind us was almost forgotten in the wonder of the moment. As we sat, recovering our breath, buffeted by the ferocious wind, she got up and moved towards us, snuffled my daughter's pack and unhurriedly returned to her refuge under the trees. She lay down again, her nose on a log, and continued to study these strange creatures who had invaded her hill. I supposed if I were starving I could bring a great stick between those eyes, but it was not an easy thought to contemplate.

We were surprised to find a sealion so high up, especially considering the difficult climb. Pete explained that the females move far from the beaches to get away from the males.

The scenery of the Aucklands is spectacular, but it was the wildlife that enchanted me. In the populated areas of the world the flash of the white tail of a frightened deer, or a glimpse of a beaver foraging for food in the evening silence of a lake, is considered a close encounter with the natural world. On these remote islands sealions, penguins, albatrosses – everything was as curious about us as we were about them. Usually they were willing to stand longer watching us than we had time to observe them. They had no reason to fear human beings and it was necessary to keep this in mind at all times so that we did not interfere with their lives.

Pete explained: "If you see any wildlife, stay five meters away from it. If it's starting to fidget – you're getting too close. Even five meters is too close in some cases. On the other hand – sealions... if you're five meters away and it comes up to you, you don't have to move away. I would advise it, but you don't have to – if you don't mind rubbing noses with it."

On our way back from the west coast, Lance and Mickey had turned the *Evohe* southward to take a good look at the channel between Rose and Enderby islands. Great waves were breaking, and my heart was in my mouth for a few minutes as they discussed the possibility of going through. The *New Zealand Pilot* describes the area in the most forbidding terms: "Rose Island is situated between Enderby Island and the main island. It has narrow boat channels on either side through which the tidal streams set very strongly, the sea occasionally breaking right across." Grampa Holding's words rang my mind, and I think had they tried to do so, I would have raced for his account, waved it in front of them and screamed, "No! No!"

In the morning I was early on the move again, determined to know if there was anything there to be found to our advantage. I struck off across the point which I knew I was on and came out on the rocks right on the channel between that island and the next. Evidently as far as it was possible to go in that direction. I could therefore do nothing but thoroughly explore that point with longing eyes to the land beyond.

My chief attention was directed to the channel which was about half a mile wide, with an island perhaps an half an acre in extent in the centre. The water appeared to be about seven or eight feet deep. Here at the entrance the rocks jutted out until they were perhaps seventy yards wide at low water and were covered at times at high tide. They were very rough and solid with no boulders or other obstructions. Between me and the island the water was quiet but on the far side the sea appeared to be of a more broken nature with breakers showing, undoubtedly caused by a reef of rocks running from shore to shore. This was indication of unmistakable danger to anything trying to get in or out by that passage so that it looked impossible for a ship to go in or out that way.

Beyond these islands was the rocky shore of the Pacific Ocean and the continuation from the small bay where I had found the first indications of habitation. The shoreline was, as in all such places, covered with holes and ditches made by the action of the water on the softer rocks. With a plentiful supply of seaweed these holes were never dry and passing to seaward they were both large and deep.

In front of me there was another island [Rose Island] placed end on nearly half way down towards the Bay which end formed the channel on that side. Taking a good view of it, I saw the side next the Ocean was an abrupt bluff sloping down towards the bay. It was bare of vegetation for some distance owing to the continual spray breaking over it. The rest was covered with a thick growth of wood right down to the water's edge. I would say one third bare, one third grass, and the balance wooded. There was quite a bit of it bare on the end as well as the side. The distance between the islands was perhaps a third of a mile. As we ultimately spent several months on that Island, I shall have more to say of it later on. My object now being to try to discover anything likely to be of future service to us.

These rocks, of a basaltic nature, were much eroded at and about the waters edge. The mussels which covered these rocks were small, up to about two inches long and in the holes I found a few with blue shells, about four inches long. The mussels were a pure godsend, but I could pay attention to them later for there were more interesting things to take stock of at that time.

It was now coming on night and time to find somewhere to sleep. There wasn't much choice so I soon found a small stream of fresh water and having lit a fire was fixed for the night. Having no fear of anything on the island I spent a fairly good night.

I was again early on the move and made my way around the rocks towards the channel. Near the centre I found the rocks level, though rough, running out towards the small island. There were numerous small holes full of a sea weed of a very tangled nature which clung to the sides of the rocks and fell in bunches forming good shelter for fish, as afterward discovered. These rocks were covered in places with snails and it was agreeable surprise to find that there was also a plentiful supply of limpets at that time. It will be easily imagined what this meant to me, as shellfish appeared to be the only thing we had to depend on except the root. Having spent a good deal of time studying this place I started further to the north where I found the rocks becoming of a rougher nature with larger water holes intervening until they reached the open sea. The foregoing description should give the reader a fairly good idea of the point on that part of the island and as I shall have a lot to say about it in the future it is necessary to tell of an adventure I had one day with a sealion on one of the holes before alluded to.

I had spent several days traversing these rocks back and forth and had never seen the necessity of carrying a weapon. While trudging round the rocks one morning near the outlet, I saw the largest sealion I had yet seen. The peculiar thing was that it appeared to be asleep in the centre of a large water hole which was about fifteen yards wide and about thirty long. I thought what a nice chance there was for me to get some meat. I did not even have a stick with me and could scarcely make up my mind how to try to get it; however, it did not take me long to decide. I went back a distance to get a stick that I thought large enough to kill it. It took quite a time for my knife had by that time become blunt, but having at last secured one I started off with great confidence. There were several holes between me and the channel with very rough rocks here and there. I went on cautiously to the hole and knew that I must use the uttermost care to enter water of unknown depth. I did not mind the wetting in the least as we were used to that. The water, being well sheltered, was quite smooth and I was afraid to cause a ripple but could not avoid it altogether.

I got to within about four feet of the seal who took no notice as he was still apparently sleeping with his nose sometimes under water (as the first one had done). I was waiting for him to rise and just as I was about to strike, he gave me the surprise of my life. Up went his snout, he gave an awful snort and a plunge, knocking me on my back and with about three leaps he was out of sight in the Channel. I quite believe that he was the more frightened of the two. This was one of the largest I saw while on those Islands. To say I was disappointed is putting it mildly – I felt crushed with despair. It must be acknowledged that I was getting very weak, as well as thin and not in a fit state to put up a fight with an animal of that kind. I was now wet to the skin and had to go and light a fire to dry my clothes. This was an experience not to be forgotten in a hurry, but it became useful at a later period.

I passed many days on this point with nothing of importance turning up but had seen sufficient to satisfy myself as to the excellent prospects of the area. I had always in mind the dread of being left alone on the island and was anxious to know how my late shipmates were getting along. I was anxious to do whatever possible for them and now wanted to get them to come back with me to this place. Preparation was unnecessary, as there was really nothing to prepare. So early one morning, I started off and by taking a shorter cut through the bush did not take so long as expected.

---

[1]   Likely *Castaways of Disappointment Island.*
[2]   Department of Lands and Survey, *Management Plan for the Auckland Island Nature Reserve, 1987*: p.17.

15

dying

much reduced

AN account in the *Aberdeen Herald*, 5 August 1865, fills in some of what transpired while Holding was away:

Week by week some were dying off. One day Captain Dalgarno had been out in search of food, and returned, wet and fatigued, to camp, where some of the seamen were seated near the fire. The Captain, unwilling to disturb the men, instead of seeking to sit by the fire, walked about to keep himself warm. After a time, one of the seamen asked his neighbour to sit round a bit and let the Captain come to the fire, but received no answer. A little after, he repeated the request, but again receiving no answer, an inquiry was made, and the poor fellow sitting beside him was found to be dead.

I found most of them in about the same condition as when I had left. Needless to say they were pleased to see me and to hear the result of my experiences. I regret to say they had been very much reduced by deaths. In the interval the Steward and the two poor boys, Liddle and Lansfield, had been carried off and the house was very gloomy in consequence. They had done the best they could to bury them. Every one was becoming weaker day by day and no one could have any idea who was likely to be the next to drop off. So altogether, it was not an happy meeting.

The second mate, Mahoney, was laying as I had left him, close to the fire which had burned into the turf leaving a large hole with bricks around it. It had by now burned so deep that it could not go out. He had apparently stripped the clothes off the boys, (or all he wanted), to such an extent that it was with difficulty that he could move. I tried to persuade the others to go back with me but the only one who would leave the house was Harvey. I promised that I would come back in a few days again or send Harvey. We started off the next morning and on the way got up a few of the potatoes which we tried boiled that night (having brought along one of the tins). Unfortunately they were no use to us as they wouldn't boil soft. We continued on and then I sent Harvey back for the others, but instead of bringing them he stayed and I had, in a few more days, to go to fetch him.

We were now reduced to seven in number and our strength was failing fast for want of food, though perhaps I was in the best condition as I had been able to get more than they. It was evident to stay there meant death to all. Persuasion appeared useless to all except the Mate Smith, Harvey and Fritz, whom I prevailed upon to accompany me back to the point. As usual I promised to send the others word of anything I found to their advantage. On the following morning we four started off.

I had taken the old adze with me hoping to kill a seal with it after putting a new handle into it later. It was useless to hope to put in a hard wood handle as we could not cut that, but there was some soft wood growing; though only small, it was large enough for that purpose. I had in fact brought everything that I could which I thought would be of use, as I did not intend to go back there to live.

I had found a little shorter route by heading off the trail towards the Bay, thereby striking a small bay on the near side and going around by the rocks instead of going all the way through the bush. There was the best sight of limpets they had seen and they could help themselves. We had not gone more than two hundred yards when I, being first, saw a seal lying asleep on the rocks just in front of us. It was quite a full-grown one of the spotted variety. Throwing up both hands to caution them, I stopped to consider the best means to secure it. Not being at that time used to carrying a stick none of us had one. The shore here was strewn with boulders, so it was little trouble to find ammunition. I went gently up to the seal and dropped a stone nicely on its head and the knife was soon into it. I need scarcely say that I did not cut my fingers that time.

I then started to skin it and found that the skin had to be cut entirely off as there was no possibility of stripping any part of it off. I have before stated that we did not get more than two fur seals – this being the second and the last of that kind which we ever saw. Having cut it up, we divided it so that the three of us carried the mean, while Fritz was appor-

tioned the liver, skin and the head. We had not gone far when someone turned round and saw Fritz kneeling down on a rock with the skin and liver, devouring it like a dog.

We had not far to go to my old camp on the side of the Bay and soon had a fire and to begin cooking a steak. By the time we had had a good supper it was almost dark. Fritz was a long time coming and only found us by the light of our fire. In our position no one could upbraid him but to our surprise he had not brought any part of his load with him. I asked where was the skin and the head and he said he had left them in the bush in the corner by the shore. It was then too late to go to look for them so they had to be left until morning. Well, we made the best we could of it that night and having had the best supper we had had for some time, lay down under some shady trees. The roots were a great annoyance, being a great deal too hard, but as there was a trickling stream running under us they were thick enough to keep us out of it.

At the peep of day I was off to get the head and the skin. I did not have more than eighty yards to go and was only just in time to save some of the skin from the three or four albatrosses. They had eaten all the meat off the head and had got the skin to the water and were trying to devour that too. Say was I mad? I think I was for the head is always covered with a good thickness of meat and this we were wanting worse than any one thing. The skin was next in our requirements and this I had providentially saved by being early. We were now all nearly bare footed and the skin was wanted for foot wear.

After I had showed them around the point they felt more satisfied with the mussels and a few other things and wished they had taken my advice earlier. The Mate and myself camped in a small hollow near some water while the other two struck up some bushes quite close to us and made themselves fairly comfortable there. We had brought with us one or two of the tins before mentioned, here I would like to correct a mistake I have made as to the size of them. There were four 4 lbs. and two 6 lbs. tins.

We camped in the same place the next night and were getting our drinking water just below where the Mate and I were laying. Sometime in the middle of the night Fritz came to us with a water can and I asked him what he wanted. He said, "You called out to ask for some water, didn't you?" I said, "No, I don't want any".

He went to the hole just below us, filled the can, then returned to his camp. When he got to the entrance to the kind of wigwam they had made, something occurred which made Harvey get up and push Fritz out and he fell on his face. In the morning I found poor Fritz laying there quite cold and dead.

Harvey said that when Fritz came back he committed a nuisance and he (Harvey) gave him a push so that he fell and never moved after. We did what we could for him by taking him under a tree and covering him with boughs. Two or three days after this I sent Harvey to

   try to get the Captain's party and told him to bring them down to us. Our meat was now getting low and the Mate and myself found that Harvey had been eating some of Fritz. It may seem strange to anyone reading this that

I go into details of these things. But I must say there is not an incident in the whole narrative which is not too much impressed into my mind, that I can ever forget.

When Harvey had not returned in three or four days we became anxious and started off to fetch them. Things were now looking very bad for us all. When I got there I learned that Harvey had died, thus leaving only five of us; We were losing our members at a rapid rate.

After explanations as to the advantage of their accompanying us back, the Captain told me that that day the Carpenter and himself had been across the point to where we had first struck the bay. On their return trip the carpenter had been unable to walk and he had carried him some distance before being obliged to leave him on the point. He explained where he had left him and though I was tired and it was getting late, I started off at once to try to find the Carpenter. Crossing the point where I had killed the first seal, then taking to the rocks, I was just on the point of going up on the next point when I saw some feet hanging down towards the water. There were here some rocks forming a kind of crevice about six feet high and pointing towards the water and covered with bushes hanging very low. This made the place very dark, so, lifting the bushes, I crawled in. There, sure enough, lay the poor carpenter. I called out, "Carpenter", but he could not reply, though he may have known someone was speaking. The only movement I saw was a trifling move of his eyes. This was, I think, one of the most touching sights of my experience. Nothing could be done, as he was without doubt the most robust of the whole lot. How the Captain ever carried him I never could understand. So I had reluctantly to leave him, knowing that he could never move from there.

When I got back to the house, just as it was getting dark, the Second Mate ordered me to go and get him some roots. I said: "No, Mahoney, you are as able to get them as I am. You have made up your mind not to try as long as others will get them for you. I have been travelling back and forth for weeks to try to help you while you are laying on your back against the fire." With that he picked up his knife and threatened to run it through me. This was too much for me. Being near the fire where he had the bricks, I picked up one and told him if he did not drop the knife I would drop him. He did not want telling twice either and he was very quiet after that. I was only young but had had some rather rough experience. I don't think I could be cowed by an Irish New York bully at that time.

We were now reduced to Four. In the morning I started off to have a last look at the Carpenter but on reaching the spot found, to my horror, that he had disappeared. The tide had gone over him and carried him into deep water. I then went back with a heavy heart to tell the others. There was now nothing to detain us here. As Mahoney would not go, (and I don't think anyone cared for that), we were soon on the way to the point where I had tried so hard to get them.

According to Smith, "Holding returned [to the point] the day after and said that the captain and second mate would join us soon, but that at present the second mate was unable to walk, as he had a very bad boil on his leg; the captain was to stay with him until he was better... After a time the captain joined us, and said that Mahoney, who was not yet recovered, was to follow him as soon as his leg got better."[1]

The inconsistency between the two accounts of the abandonment of Mahoney is interesting, for with this one exception the two accounts are very close in all respects. The captain, by leaving a man unable to walk and thus search for food, would have essentially condemned Mahoney to death – but is it likely that he would have gone to the point alone over an unknown path without a guide at a later time?

[1]   Smith, A. *The Castaways*, See Appendix Three.

much better today

fish

We made good progress as I had gone over the ground so many times. Nothing of importance was seen on the way except roots and limpets which were plentiful and of good size. We enjoyed them greatly after roasting and had all we wanted.

So that we would have a better view of the water front I had chosen another place to camp in a good sheltered position right opposite the channel and the island. The whole distance from the bay to the open water would not be more than an half mile, thus we could at any time have a good view of the ocean by going that distance to a rocky promontory, or to the Bay from the point. This ultimately proved a boon to us. On our left was thick green bush and on our right low bush for a little way, then a patch of long flaggy grass. This indicated to me that that place had been used for the same purpose as that around the house.

I feel sure they never regretted the change. Nothing could be done that day except to make a place to sleep. And here I may say, we might often have spent more time on our camp, with advantage to ourselves.

During my hunting round the rocks and finding every hole edged with seaweed, I thought there ought to be something there besides mussels and soon discovered that they were the refuge of fish. Having previously stated that we had found some fencing wire, I may now say to what use it was put. It was not long before I had made a spear in this way: I heated one end, laid it over a stone and struck it with another – bending it about half an inch. I then burned a hole in a stick which I had cut about six feet long. Leaving about nine inches clear, bound the wire on with some strips of a handkerchief. This I used to turn the weeds gently aside and spear the fish as they lay there. I could often spear four or five in a short time.

Having now a good supply of meat and some skins to spare, I thought I would cut up some of the latter and make a net of them, for the spear was rather slow work. I got some twigs and made an hoop and with netting cut from the seal skin made a kind of bag net or creel, such as

I had heard they caught lobsters in. In this I tied some of the seals entrails, went out to some rocks where I could reach over the water and lowered it down in about twenty feet of water. I caught a few but found it a very slow process so did not consider it a success.

As I had plenty of time I set out to make another type which we used to call a bow net when my father was gamekeeper for the Duke of Manchester. I placed four hoops about one foot apart and stretched them apart with three sticks. An entrance at each end narrowed down to about one third and each is triced to the opposite end, so that the fish going in at the wide opening squeeze through into the surrounding space and seldom find their way out. This I charged with offal. (I may here explain, that being the son of an English Gamekeeper that I was taught to do some netting by the time I could talk.) This was not a success either as the fish would try to eat the skin and could get at the offal from the outside.

* * * *

It is surprising that Holding does not mention one 'peculiarity' of some species of fish caught off the island which was mentioned by several visitors to the islands. D'Urville gives a very vivid account of what he encountered:

Their flesh was infested with strings of very fine worms running in every direction and giving a marbled effect to the fish. For the first few days that we were in harbour here, our sailors thought that these worms were merely veins and they ate them freely, without ever suffering any ill-effects. But later on, when they saw that most of the officers turned with disgust from this infected fish, they too preferred to eat their ordinary rations of salted fish and from that time fishing was to some extent given up. It is quite clear that this peculiarity was characteristic of almost all the fish caught off this coast. The most vigorous of them, no less than the others, were riddled with huge worms of astonishing thickness and length. I have often had fish cut open while they were still alive, just before they were cooked, and I have seen for myself how almost all of them showed the same peculiarity, of whatever size and variety they might be."[1]

These things helped to fill many of what otherwise would have been weary hours while sitting on the high rocks overlooking the ocean hoping to get sight of a vessel, by giving me some occupation as well as enjoyment – to say nothing of food. We were now quite into the spring and there were many pleasant days as far as the weather was concerned, but over all always the mortifying thought of our exiled position. We knew seal had been up in the bush but had been unable to see them. We were now getting mussels, limpets and fish as well as roots. These latter we continued to eat no matter what else we had but as previously stated the roots were of a very binding nature. We found that the limpets acted in the contrary direction to an alarming extent.

Our one wish now was to get some seal meat; then we thought we might manage to pull through until someone visited the islands. The captain often asked me if I thought we would ever be relieved and I always said yes, but that I thought we might be there for three years but not more.

We were, by the Captain's reckoning, into August and the days were stretching out. I felt anxious to know the fate of Mahoney, so I believe it was the Eleventh that I started off to ascertain. The going was fairly good and I was not long on the way.

He lay just as we had left him; I don't think he ever tried to move. I find I have overlooked the finding some bottles and we had left some of them filled with water by the side of him. They were still there but I don't know if they had been touched. The bricks round his fire had partly fallen in the hole by his feet. His body was too much decomposed to even touch. I took one of the two roof slates which we had found and with the point of my blunt knife, scratched,

"JAMES MAHONEY. WRECKED. WITH THE SHIP INVERCAULD

MAY 10th 64."

and this I tried to sign Robt. Holding. I placed this by the side of the corpse near his head and left him with a sorrowing heart, although this had not been unexpected. Now only the three of us were left out of Nineteen which managed to land.

I might here say that that slate was taken to Melbourne by Captain Musgrave on his discovery of the body and placed in the museum there near the end of '65 and may still be there with other relics. I don't think the writing was ever deciphered, as it was stated, "That it consisted of Hereogliphical [sic] zigzags, but they did appear to have made out the name Mahoney.

* * * *

We have tried to track down this slate, but to no avail. At best, it is probably sitting somewhere in the depths of a museum, uncatalogued; at worse, it has been tossed out as simply an old piece of rock. The identity of the 'mystery body' fired the imagination of those who heard Musgrave's account after his return to rescue his men. He had described what they found when they stopped at the Hardwicke settlement:

...in the midst of a coppice, at no great distance from the shore, we saw a number of wrecked and shattered huts. Each stands, or stood, in the centre of a small enclosure, designed for a garden, and surrounded by a ruined palisade. In traversing these ruins, we arrived in front of a hut less dilapidated than the others; the thatched roof appeared to have fallen in a little. Hardly had we entered it before we recoiled with fright, or rather with horror. In the corner of the interior lay a dead body. It was that of a man, who must have been dead for some months.

Overcoming our first emotion of repugnance, we approached it. It was lying on a platform of planks, evidently procured from the hull of a ship; these were supported on logs, and covered with a layer of moss. The arms stretched by the side of the body, and the fingers of the hands straight and untwisted, were indications of a peaceful and apparently resigned departure. One leg hung a little out of the bed, the other was extended full length upon it. A shoe was upon the left foot; the right, probably wounded, was wrapped up in a bandage. The dress was that of a sailor; more over, several garments, one of which was an oilcloth overcoat, were thrown upon the body to serve as coverlets. On the ground, near the bed, lay a small heap of limpet shells; and still nearer, a couple of glass bottles, one full of fresh water, the other empty.

Finally on the bed itself, within reach of one of the hands, we found a slate, on which a few lines had been written. Upon it was the nail with which they had been traced. We attempted to decipher the writing, but could not succeed; the rains and wind had rendered it illegible; or perhaps it had been scrawled by the trembling hand of a dying man. A single word was tolerably plain, the name of James, forming a part of the signature; the other, completing it, answered in the form and number of the strokes, to that of Rigth, but we could not say so with any certainty. I brought away the slate and will show it to you.[2]

How did this corpse come there? We could answer the question only by vague conjectures. That a ship had been wrecked at Port Ross, or in the neighbourhood, was hardly doubtful. Perhaps the crew, with the exception of this one man, had been drowned: this would account for the clothes, which the survivor had collected and heaped upon his bed for the sake of warmth. Or several unfortunate castaways might have reached the shore and, finding no means of sustenance at Port Ross, had advanced into the island: the smoke which we had perceived, or which we thought we had perceived, on the mountain side, might be, perhaps, a sign of their presence. One of them, wounded in the foot, had remained behind alone; he had taken refuge in one of the huts, whose roof was not yet shattered; incapable of hunting the seals, he had lived for a while on shell fish, and at length had perished of hunger.

Looking at this poor abandoned corpse, we felt a deep compassion. Our thoughts naturally dwelt on what might have been our own fate. We were unwilling to leave it unburied. Next day we dug a grave and reverently interred it; and after saying a few prayers over its last resting place, we planted there a wooden cross.[3]

But Mahoney was not left to rest. Coincidentally, only three days later the *Southland* arrived in Port Ross, also to search for shipwrecked mariners. Baker, a surveyor on the ship, explains:

The captain had a coffin constructed, and the skeleton was exhumed, enclosed in it, and re-buried in the cemetery which had been used when the Enderby Settlement existed. Dr. Moncton was of the opinion that the man had been dead at least 12 months and his clothes showed that he was not an ordinary seaman. More than this we could not gather: there were no marks of any description, no writing in the hut, nothing to suggest how the man had come there, how long he had lived in this deserted spot, what he had done while there, or how he had met his end. The mystery remained unsolved.[4]

The 'special reporter' from the Christchurch paper let his imagination run wild as he recounted Musgrave and Raynal's sad discovery. After referring to the 18 [sic] who "went singly into the hills" he continues:

...Whilst that one returned wanderer from the strange quest lay dead or dying in the hut, another fearful tragedy was being enacted. Another fine ship went ashore on the south part of the same island, and another party of sailors escaped the sea to find themselves imprisoned on desolate land. Some of the sailors, headed by the Captain, set off to explore the island, and managed at last to reach the hut built by the *Invercauld* people. The Captain entered it first, and saw the dead body of a man lying on the floor, covered by a torn oilskin coat. A piece of paper was lying by his side, and as soon as the captain saw it he exclaimed, "Why, it is my old friend who taught me navigation and seamanship", for the piece of paper was signed Maloney, [sic] second mate of the Invercaul [sic], told the captain that it was indeed his friend.[5]

Fortunately, this version never entered the recognised historical record; however, well-known books have confused the facts. W. Jeffery in *Century of Our Sea History*, 1901 wrote:

The schooner sent to the rescue of the *Grafton's* crew, the *Flying Scud*, while she was in the group, came across the skeleton of a man, evidently one of the *Invercauld's* crew, who had wandered from his fellows, and fallen, exhausted, to die on the rocks, but there was nothing by which he could be identified.[6]

I returned to the point the same evening and stated how I had found things there. The Reader may be sure we did not feel in the best of spirits. These things had been coming upon us thick and fast for the last few weeks and our lives still hung in the balance. The fact of the limpets having spawned and left the rocks all in one night did not lighten our burden. Our nights became very restless, with all sorts of dreams and our dispositions towards each other none too good.

About this time we did occasionally see a seal swimming round and soon had the good fortune to kill one. As well as the meat, we found a use for other parts. We had, as I have stated, made some low moccasins and while going through the holes by the sea I had noticed that the fish bit at them. This opened up another idea. I cut about six feet of the entrails of the sealion and tied it round one leg then went into the water holes about up to my knees; the fish would take hold of the end and gobble up all they could, then begin to twist round, apparently trying to twist it off. That was my time to grab them. When they found themselves out of the water they would disgorge so that I could then throw them out. In this way I have often taken two at once.

I used to give the others some fish when I caught them, which was almost as often as I liked to go for them, but I found that by being continually in the water my legs swelled to quite double their size. Ultimately I was compelled to adopt some other plan. I had cut up some of the seal skin into strips about the size of boot laces to use for carrying the fish, but found that very slippery when wet. I well remember that on one occasion, when I had caught twenty-six I had to pass the mate who was on the rocks gathering mussels. I called out to him and asked him to help me carry them, as they kept on falling. This he refused to do. The consequence was that I told him he should have none of them and if he wanted fish he would have to catch them himself in future. This, of course, caused an estrangement between us all.

The foregoing and a few other incidents which occurred often caused dissension between us and at last it culminated in disruption. At last they decided to make a hut for themselves not more than sixty yards away but that was far enough to keep us apart. I cannot say it was of any disadvantage to me, for the place we had put up to sleep in was not large enough for us and they wouldn't try to improve it. I now had more room to lay down and thereby got better rest. As they were seldom out I did not see them very often. It was a disgraceful state of affairs in our position, but it shows how easily people can become divided under such circumstances. Whether it was their fault or mine the reader must be the judge and as it turned out we didn't court each other's company for several weeks.

Although we now had lots of meat, we did not relax our efforts to get a sealion when we had the chance. This now appeared to be the best season for them as they were frequent visitors now and for sometime. It would tire the reader for me to try to recapitulate one half the adventures I had with seals, so will content myself with giving a few of them.

I had put a handle in the old adze and thought I had just the implement to make quick work of killing a seal. Owing to the frequency of their visits I seldom went out without it but soon found by practice that it was not the best article to use as the weight was too much at one end. More than once it was the cause of me losing one as it flew out of my hand when I struck. Besides, the blow falling in one place did not have the effect that one would have were it distributed over a larger area. To my mind a baseball bat would be an ideal instrument for the job.

Where we were camped the distance from water to the bush was about eight or nine feet and was strewn with boulders of various sizes worn smooth by the action of the water. One day, while passing along there, I saw a very large sealion and was determined to take a chance at him. Once in close quarters these animals don't often try to get away but stand with wide open mouths or come and face you. At that time I had but little experience with them and being by myself the look of it almost struck me with terror – but hunger is a sharp thorn. So making straight to him, intending to hit him on the head, I got to what I thought was striking distance and made a lunge. My foot slipped on the greasy rocks and though I did hit it, the blow had little or no effect for the adze flew out of my hand and he was in the water before I could recover.

One morning the mate and myself were on the point near the bay when a sealion passed us, swimming only a short distance from the shore. The sun was shining brightly and I watched it go across the bay and climb out. I said to Smith "It has gone up into the bush. Let us go and try to get it." To my disgust he positively refused. I told him if he would not go, I would by myself and started off as hard as I could go round by the rocks. It was impossible to hurry for the large blocky stones were slippery.

Having the place marked, I had no trouble in locating the place and I at once found his trail. The bush was fairly open here but in some places it was with difficulty that I was able to follow him, often having only a leaf turned upside down showing the wet side instead of the dry or a blade of grass knocked down to guide me.

In this way I kept on for perhaps three hundred yards when I heard a grunt and, knowing the significance of that, went more cautiously until at last I spotted my prey. I got a bush between us so that it wouldn't see me and then made a rush for it. As was their custom he turned on me with open mouth. I had by this time become aware that it was next to impossible to kill them by hitting them above the eyes, owing to the amount of flesh contained on their heads, but that one good blow anywhere between their nose and eyes was generally effective. I had no difficulty in stunning him with the first blow and as he fell over, much to my surprise, another one jumped out from alongside of him started on the jump down the hill towards the water. I did not stop to see if the first was killed but went after the second. Getting in front of it I served that one the same, then went back to complete the job on the first. In place of one we had two. What a glorious feed we would now be able to have, as well as a much needed little rest for a few days. They are not always so certain of being caught, for I remember upon one occasion meeting one near the same place. He had bolted down the hill before I could get in front of him and plunged into a ditch overgrown with small bushes. I never saw him again.

It was only the work of a few minutes to open up the skin and obtain some steak. On going back to the Captain and Mate I tried to shame the latter for his laziness but it didn't appear to have much effect. They were, however, glad to accompany me the following morning to skin and bring the seals back to camp.

These two gave us a good supply of meat for a long time as well as some extra skins. The worst of it was that we had no knowledge as to the tanning of them. The plague of flies was one of our greatest troubles for there were an enormous number on these Islands. They consisted of the house fly, the blow fly, and the bluebottle. Each of which would drop live maggots as they ran about on everything, even a damp board on which meat or fish had been laid or cut up. When we stretched out a skin to dry they would collect on it and leave large bunches of the pests.

One night shortly after this, while laying half asleep, I heard a noise outside my teepee. I lay listening some little time but could not make out what it was. As it was getting closer, I rose to have a look. I may here explain that by continual use we had made quite a trail and it was along this that the noise came. It was little more than heavy breathing, but would cease for a few seconds then start again. At last it stopped right opposite my door, which I had built of brush so that I could lift it in and out of position.

The moon was shining bright but I could see nothing but a black shining object, which I concluded was a sealion. My sole thought was how I was to get it for the adze was just outside if I tried to get it I would lose the chance. There was only the one thing to do and that was to let it move again, then push the door to one side and secure the adze. This is what I did and was surprised to find that he had entirely given me the slip and could not be seen anywhere, although I hunted well for it. There was, however, the satisfaction of knowing that we were in a good position for them and that they were cruising about more than they had been doing of late. I told the others of my failure, though that was poor consolation.

One day while going round by the rocks, which were very rough at this point, I caught sight of one of the biggest brutes that I had ever seen. The water was within about nine feet from the bush and the shoreline was rather steep which made it a very difficult place to kill one and having only the adze with me, made it still worse. However, I had to try as he didn't seem to care to move, so I went boldly up to him and as he opened his great mouth I made a lunge at him hard. I think I struck him in the left eye but at the same time my foot slipped and my left elbow went directly into his mouth. He gave just one grunt and disappeared into the water. I would here say that in striking them with a swinging blow from the right I invariably caught them in or near the left eye and most of them that got away went with one eye. I believe I had the honour to finish that big fellow a little time after in the following manner.

There was, just below my camp, a bunch of stumps of ferns near the water. These were about two feet high with a few leaves on the tops and amongst them some tree ferns about seven feet or more long. We never knew how they got there or for what purpose. One day I saw, just above them, a very large sealion and called to the others to come and assist me to get it. It was always policy to get between them and the water, so we all went round on the rocks. It was then decided that they should stay on the rocks in case he bolted, while I went up to him with my club. As I got near he kept turning to the left and after running round him several times to try to get a good blow, I noticed that he was blind in the left eye. So instead of following him, I turned and met him and in this way I was able to strike him. Unfortunately, it did not have the desired effect and he began to try to get to the water. Well, I pounded that great brute until

I was tired, but he still kept on going until he got to the ferns. It was plain to see that he was getting weaker, as well as myself, and I am sure my blows did not have the effect that they ought to have had for it was cruel the punishment he was getting. However, I was now confident that I could finish him on the rocks as his other eye was also gone. I purposely let him go down then met him with one on the nose and soon had the knife in him. This we did after one of the toughest jobs we had had and he was one of the largest we got at any time.

While we were well supplied with meat, I had made up my mind to try to build myself a better place to live in and chose the spot where the long grass was growing and here I built myself a nice grass hut. I set up some stout sticks and using others for wall plates and rafters, tied these together with seal skin strips, and covered them with the long grass. I made the roof the same way for a place about six feet by eight. I then made a door to close up by lifting it in and out and built a bunk. In this I lived comfortably as long as we were on the mainland and taking things altogether considered myself fairly well fixed.[7]

Coming out one morning I saw a cow seal amongst the long grass near my shack, so taking my club went for it. She started for the water at once but I got there first and killed it on the rocks. The others had reached me by the time I started to skin it, for this was always my job. I was about half done when the other two, who were sitting near me, called my attention to another one swimming round by the shore. I told them to lay down and did so myself. When the sealion got right opposite, he came straight out on the rocks within two or three feet of me. Having my club handy, it was but the work of a moment to give him one on the nose. This laid him out, of course, so instead of one, we had two to skin.

I had the one finished and partly the other when they called out again, "Here comes another!" We all lay down, me with the first one's skin over me and club again in my hand. This one actually came up and smelt near me. I threw off the skin and he went down as quick as the others. Thus we had the three laying side by side in a very short time. There was now no cause for worry about food for some time as we had one whole seal each to fall back upon. We knew that we couldn't consume the whole of them while they were fit to eat but still the skins were valuable to us. We had had meat now for some time, but latterly it had not been good owing to the flies. In fact it was almost unbearable. But I need not trouble the reader more than to say, we had much experience of that kind.

The foregoing will give some idea of what might have been if they had taken my advice in the latter part of May or early part of June. It was now late October.

---

[1]   D'Urville, D., *The Voyage of the Astrolabe*: p.10.
[2]   No name on crew manifest is anything like 'Rigth'.
[3]   Raynal, F.E., *Wrecked on a Reef: or, Twenty Months among the Auckland Isles*: p.317.
[4]   Baker, J.H., *A Surveyor in New Zealand, 1857-1896*: p.81.
[5]   Musgrave did not know Mahoney. Indeed it was they who placed the "Unknown" marker.
[6]   Jeffery, W., *Century of Our Sea History*: p.312.
[7]   Ocean Point, is now known as Lindley Point. The *Southland* party found the remains of this hut in 1865.

weather modesty

uild boat

The weather was becoming warmer and the sea gulls were busy looking for breeding quarters. I would often sit for hours taking pleasure in watching them. They always appeared to choose their own rock to sit upon and if they were a mile or more away when another settled there, they would come screeching back and the stranger had to clear out quickly. They would bring in a small crab or shellfish, fly up in the air and drop it on the rocks to break it, have their feed and go after more. I shall later on have something to say regarding these birds. Meantime we will introduce another subject. I had always had the other island on my mind and was anxious to go across the channel, especially after having seen a very large seal near the top. The advantages would be great if we could get over there: firstly, there would likely be more seal visiting the other lands; secondly, we would be able to prospect the island and thirdly, I thought the further we got seaward the more likelihood there was of seeing a ship for here we were almost confined to a view of the bay. I began to formulate a plan. I felt sure that we could build a kind of canoe with wicker work covered with skins for we had, I thought, enough for the purpose. I broached the subject to the others and after the matter had been thoroughly thrashed out it was decided to start upon the adventure at once.

During my travels, I had seen some small wattles which I thought would answer our purpose. So one fine morning we started off round the bay, went a little way into the bush and found all we were likely to require.

While they cut and trimmed I started in on my first experience in basket making. We laid out a frame of undetermined size, but which might have been eight feet long by perhaps three feet wide and about sixteen inches deep which we covered with five skins. The skins were green, that is, with the hair on, which we turned inside. We were not long making the wicker work, but the lacing of the skins caused us some trouble for we had to cut some hide for the purpose and make holes with an awl made of hardwood. Considering that none of us had had any experience along these lines, we did very well. I am not now ashamed to say we had not the common sense to steam the stuff, or we might of made a better job of it. I had at that time never seen a canoe but I can now say it resembled an Indian Canoe in shape except that it was level from stem to stern.

One day, while we were working, we heard a slight noise and on looking for the cause saw a nice cow seal a good distance from the water so that it was little trouble to kill. We soon had three nice steaks off the neck, which we roasted on sticks over the fire. I believe we enjoyed that more than any we had previously eaten.

Another time, I had just got onto the rocks and was in time to see two sealions, one near the bush and the other coming out of the water. They went for each other at once and fought a bitter battle for two or three minutes – up and down in and out of the water, with mouth and fins, until one got tired of it. He plunged into the water and went across the bay at a terrific pace and was out of sight in about a minute. I had often seen them with lumps of skin torn from them and did not wonder at it any more for while it lasted the fight was really terrific.

I believe it was on the fifth day of November when we got the canoe finished at last and carried it to the water. The Captain christened it but the name he gave I have long since forgotten. Since the job of manning it devolved upon me, I cut a pole and although we hadn't got the skins on as I wished, we soon had it in the water and was soon on my way. Much to my surprise there were a large number of lobsters near the surface feeding on small shell fish. I caught three and need scarcely say we were all delighted at my success on that first trip. Of course catching them in such a frail craft was not all fun. We saw that they were very plentiful lower down, thus it was necessary to try to devise some other means of catching them.

I first thought of a hooked stick but this would not do, for they were pretty lively and would slip off. Taking the wire, which I had been using for a fish spear, I fashioned a hook, somewhat resembling a shepherd's crook, but on a smaller scale. In about two hours had caught forty-eight, most of them over one pound in weight. I could have caught more but being that they were all alive, that was as many as I could keep in my boat. It was amusing the job we had with them on shore. This gave us an agreeable change of diet for a time and looked like very good business but it was too good to last for I am sad to say the next time that I went out they were very scarce and I could only get three or four which were in very poor condition. It had become their spawning time and in a day or two they had vanished.

The seals, too, did not appear so often and in consequence I began again to cast longing eyes at the island opposite. Shortly after I decided to go over or die in the attempt. I began by fixing up my boat to the best of my ability. The first thing to do was to make a pair of oars. I have already stated that there were two kinds of wood on these islands. The one too hard to work and the other soft and not running very large. This type grew in clusters, frequently straight with branches nearly upright and a long wire-like leaf.[1] This is what I must use. I had spent a good deal of time trying to sharpen the old adze with a brick but could do a little with it so finished the rest with my knife. In this way I managed to make a pair about four feet long. Then there was the question of supports as the frame was very delicate. I drove in a pin on each side and slipped them through a piece of the hide. I was ready for the trip.

I started off, telling the Captain and Mate that I would see what the island was like before I attempted to return and thought that I would be away two or three days. Having now satisfied myself that my boat would carry me across the channel, I was anxious to see what the other island was like.

I did not trouble to take anything with me to eat, as I expected to be able to get it over there, so only took my club. As before stated there was an island in about the centre of the channel and the tide was rather rapid there at times. I chose the time when it had begun to run out (or what is called on the ebb) and was not long discovering that I was wrong for once. I must acknowledge that I made a terrible mistake and one that nearly cost me my life. I kept well into the bay and got over the first channel and by the island all right.[2] It was when I got into the middle of the second channel that I found that the tide was running a great deal stronger to seaward than I had expected and I had not gone far enough into the bay. I began to put a little more power on the oars when to my horror one of them snapped off short.

What to do now? I was being taken out to sea quite fast and the further I went the stronger was the tide. There wasn't a minute to lose. The wind was blowing in my favour and having an extra skin with me I thought that might carry me across if I held it as a sail. That idea was a rank failure. I tried to use the remaining oar as a pole to propel the boat and edge in for the shore but I was still being carried out at an alarming rate and my boat was now half full of water. I was none too soon in doing this as the water appeared to get shallower and the tide stronger all the time. At last, after the uttermost exertions I got amongst the loose rocks; my boat struck a rock and I was saved. If I had gone ten yards further I think it would have been all off with me; as it was, I was up to my waist in water. Now it was a matter of saving my boat, for without it I could not get back, in which case I would be alone. After recovering my senses a little I dragged it up a little at a time, emptying the water out of it as I went, and at last had the satisfaction of seeing it above water. It was quite a job to get it into suitable position to work at.

To get my boat around the point I had to take it apart. I stripped the skins from it and carried first the basket then the skins. It did not take me long to find a stick to make a new oar and soon had her in the water again, but in the bay this time. Feeling none the worse for my exertions, I paddled round by the lee shore where the water was smooth as glass and made good progress even with my primitive outfit. I had gone but a short distance when I caught sight of three very small ducks only about the size of a quail.[3] Finding that I could get quite close to them, landed, picked up some stones and soon got one or perhaps two, I forget which. As I passed round I saw others, always two ducks and a drake. I might here mention that at a later period I saw many in three different sizes, they were all brown and the two largest sizes were very pretty about the head.

I found that this island[4] was about an half mile wide and about a mile long with high bluffs on the seaward side, falling off towards the bay. The higher land was bare, but the lower half was timbered with the hard wood which I have since learned is called the fucher [sic] tree.[5] It did not take me long to reach the bare rocky point which was perhaps twenty feet high. I was bound to see all that was to be seen, so passed on until I came to some bare land which extended to the next channel. Here I landed and was soon rewarded by the sight of rabbits, so it may be easily imagined that I considered myself pretty well rewarded for my trouble. I had not been there long when I saw a sealion land and letting it get well up, soon had the satisfaction of getting it. It was now getting late so I made the best of it for the night.

The next day I was astir early and soon decided that this was the best place for us. There were no roots and very little herbage on the island as the rabbits, being very numerous,[6] had no doubt cleaned them out. In fact, they were just about starved to death. The low vegetation meant there was no trouble in getting about (such as we had had to encounter on the main island). The soil was peat and comparatively dry so that it was, in all respects, a more desirable place to live. That day I had the satisfaction of killing a nice young seal. Poor little fellow, it did look like a shame to kill it.

Having seen about all that was to be seen I made up my mind to go back on the morrow to see if something could not be done to get the other two over. Of course it was impossible to take them over in my basket boat, especially as she was leaking a good deal owing to the skins stretching. Before going back I took another look around. Another seal came up from the sea and while it was crossing a bare hill I went and killed it – thus making three in three days. The young one I took with me and intending to go back for the others before long, I just left them there. This time I studied the tide so not to be caught again and keeping well into the bay I arrived back without difficulty. The others were much encouraged by what I had to tell.

The next morning I returned to Rabbit Island and got there just in time, for three albatrosses were beginning to devour one of my seals. Some will tell that they cannot start to fly on land – well they did and they did not wait for my club either! They were on a hill where nothing was growing and when they started to run the wind got under their wings and the rest was easy.

I got back that same night and we discussed the next question – how were they to get over? I proposed that we go back to the house, pull it down and try to build a boat out of the boards. How could we build a boat without tools? I said that I thought we could and after some explanations it was decided to try.

It is impossible for me at this length of time to attempt to state accurately the length of time which might have elapsed between the different events recorded. So I shall not attempt to do so! Suffice it to say that we went and that that was the first time they had been there since before Mahoney died. When we arrived we took a long look at what was left of him. There appeared nothing but clothes and bones, all in such a state that it was impossible to move it. All we could do was to pull down the place on top of him and nothing could possibly hurt him then.

We were not long in demolishing the place for they had burned a lot of it. Using a round stone held in the palm of the hand, we drove out the nails by forcing them through with another nail and then straight-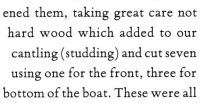ened them, taking great care not to lose one. The boards were of hard wood which added to our trouble. I now got some of the cantling (studding) and cut seven lengths about two feet long – using one for the front, three for each side, then three for the bottom of the boat. These were all cut with the old adze with only half the face. This was the only tool we had as the old axe had a bursted [sic] eye and could not be of use. In this way we built our boat. The size might be near eight feet long, thirty inches wide and twenty-four deep.

While we were back at the settlement we looked up everything likely to be of use to us – even the old spade for which I later found a use on Rabbit Island. The new boat did not consume half the boards so before going across the channel, we thought to take some of them with us. I told the other two (who never having hurt themselves with work) that I expected them to fetch some down and tow them across. This was agreed upon as they were now just as anxious to get across as myself. The days being long now, I thought that by starting early, they could easily do the trip in the day.

They started off one morning and were soon round the point and out of sight. When darkness set in and they had not returned I began to get anxious about them. I saw nothing of them until morning and when they returned I inquired what kept them and asked where were the boards. They told me that in coming back they had caught the raft in some weeds and were obliged to leave it there but had tied the raft to them. I then inquired as to where that was and they said, "out in the channel." I immediately went to look for it, but it was not there. They had evidently tied it to some of the light weeds such as the crabs had been feeding on and when the tide rose it had broken adrift and floated away. This was annoying enough but there was worse to follow.

I now wanted to know where they had left the boat. They said "in the bay, in the corner where Fritz had left the seal skin and head". I said, "did you pull it well up?", and they said they did. So without delay I started off to get it but my fears were soon realized for the tide had carried that away also. This was near the cause of a fatal split between us, as anyone may easily imagine.

* * * *

Not surprisingly, Smith describes the events somewhat differently:

Soon after building her we went and collected all the wood left of the house, and rafted it near to the place where we were living, as we intended to build a sort of house with it, and make ourselves as comfortable as the circumstances would permit. But in this we were disappointed, for during the night a heavy gale came on which brought a great swell on to the beach, and carried off our boat before morning. We had no other help left now, but to commence and build another boat with the remaining wood.[7]

Incredible as it may seem, detailed weather records exist for much of the time the *Invercauld* survivors were on the islands, for Musgrave had rescued a barometer and a thermometer from the *Grafton*. From his description, Holding, Smith and Dalgarno had to work under difficult conditions:

January 10-25: a rotary gale continued without ceasing – rained heavily constantly.
January 25-31: three clear days.
March 3-12: some of the finest weather since we came.
March 17 and 18: hurricane – heaviest experienced.
March 21: wet and boisterous for the rest of the month.
April 2: rained heavily since noon.
April 3-16: incessant gale with constant snow, hail and pelting rain.[8]

Well, recognizing the fact that our interests were of a mutual nature, we became a little reconciled and decided to try to build another boat. We lost no time in getting at it as this was the most pressing matter before us. We found we had sufficient boards and having had some experience it's likely that when finished it was better than the one we had lost. We still had some boards left and decided to tow them down also. Be sure that I took no chances with them this time! This will account I think for the fact that there was little to be seen when Mahoney's body was discovered by the Grafton Party.

Drawn by Holding's memories, we left our friendly sealion and the magnificent view from Dea's Head to continue along the route which he had travelled so many times. The afternoon shadows were lengthening as we finally stood on the easternmost point of the main island. Nearby, somewhere under the tussocks, was Fritz's final resting place. So many men had died; it is ironic that Mahoney is the only one with a marked grave. Now that locations of the deaths of the others are known, it seems only right that there should be a memorial to them as well on these islands.

I felt that I had been here before, for my great-grandfather's descriptions were so accurate that there is nothing which I can add. The seaweed-filled holes, the limpets clinging to the basaltic rocks, the waves pounding through the channel, the islands – these were all just as he had described. One or two young bull sealions were swimming around, bobbing in and out of the deep holes; one fellow came in with great snorts and roars before turning back out to sea as a rival took his place. It was almost as if they too had read his account and were setting the stage for us.

Wrapped in our own thoughts about the man whose story had brought us here, we stood on the dark, ragged rocks. The waves crashed and ripped through the gut between us and Friday Island and beyond to Rose. I could only marvel at the bravery of the young man who took a leaky craft made of sealskins and wattles across those fearsome waters. Incredible what the human spirit can do when survival is at stake.

This was really the first location we had been to that was unique to the *Invercauld* story. We had not actually stood at the site of the wreck and although the tops were extraordinary for their wild and bleak appearance, we did not know precisely where the men had been. But here on this desolate shore, Dalgarno, Smith and Holding had lived, planned, and managed to survive. Apart from taking photographs and scooping some sand into a film container, there was nothing we could do to mark the site. I felt a melancholy sense of anti-climax as we turned and headed back for the pick-up site.

It was here too that we did something which we realised at the time was foolish, but fatigue and perhaps a sense of security had set in. The dinghy arrived and to save the crew an extra hour of driving time we all piled in. We were grossly overloaded and even though we were creeping closely along the shoreline, it was uncomfortable. This was the only time on the whole expedition where there was an unspoken sense of contention.

How easy the trip back along that shore where we had slogged throughout the day. The beach masters raised their heads at the unaccustomed sound of a motor slicing through the silence. Dea's Head, with its columnar rocks, was more impressive, if anything, from below. Yellow-eyed penguins and shags shrugged as we went by. We were relieved when we safely reached the *Evohe*; a quick shower, a hot meal, friendship in the warmth of the saloon and bed. How different from the shelter of the men of the past who could only suffer and wait.

*   *   *   *

Unlike the punt built by the *Derry Castle* survivors, the *Invercauld* coracle has disappeared into the mists of history. It was mentioned only by the *Southland* party:

...the frame of a boat formed of woven sticks and lashed with seal skins and close at hand a thatched hut, about nine feet square, evidently built by men who had no tools, with a track leading from it to an eminence overlooking the sea on which a pole with a bundle of white grass had been erected as a signal. The various indications led to the conclusion that at least 12 months must have elapsed since the wreck of which these are the evidence. It is much to be regretted that nothing has been discovered affording any degree of certainty as to either the vessel lost or the fate of her crew. Having recently received the narratives by the captain and mate of the Invercauld on this coast, we incline to the belief that the traces which our reporter mentions tally most strikingly with their statements, and in all probability are the traces of this wreck.[9]

Captain Greig wrote: "On an eminence a little to the westward, which commanded an extensive view of the ocean to the northward, lay prostrate a resemblance to a cross, probably put up originally for a signal staff. On the beach a little distance from the point, was the wreck of a small boat, made of small sticks interlacing each other like wicker work, and fastened with strips of seal hide. This last mentioned material had probably been used for covering the outside of it. ...About two miles down the bay, on the south side, we found an old ship's boat, square sterned, about eighteen feet long, entirely weather bleached, and overgrown with moss inside. No name on it."[10]

John Baker, the surveyor on this voyage, also mentioned this discovery and identified the location of the sealskin boat as Ocean Point on the east side of the harbour.[11]

It was only a short time before we started to cross the channel and by keeping well into the bay we arrived without mishap. There was good shelter everywhere on the one side of the island – that is to say, the side next the bay. For our camp we chose a place about thirty yards from the water with a good tree as background. We put the boat a short distance away, for the rocks just here were perhaps four feet high at low water and the water about the same depth. We had no shelter except the trees and it now being summer time, we were getting much rain. To protect us from drafts on one side we broke off branches to go round the bottom while the other was left open.

I was still making plans for the future, especially in the way of building a house. In our hurry to get away we had left the old spade and a few other things behind, so I proposed that the captain and mate should make the trip to fetch them and they gladly assented. They got ready and we brought the boat round. We had some good skin for a painter and the boat was fairly tight. I had, in building it, taken the precaution to nail a scantling along the bottom for stability. They had, as I have said, both saved their overcoats and would take them with them, though there appeared no reason for doing so. They got in all right but for somehow capsized about ten yards out. It was a good thing for them, I am inclined to think, that I was there and had with me a good skin rope. While they were floundering about, hampered with their coats, I threw them the line and told them not to come in without the boat or oars. I could not help laughing at them as they called out to me, "For God's sake, save me, I cannot swim", while they were trying to save their coats. Well, we eventually secured everything, got the boat out, emptied. It was but a short time before they were off again, this time without trouble. Being only a nice trip, they returned without accident.

Now to begin our house. The soil just outside the bush would have been very suitable for cutting turf had not the rabbits kept the grass down so low, but we must have that or nothing. I got them to fetch up stones for a chimney while I cut turf with the old spade. Having cut a quantity, I began to lay out our house which I figured at about eight feet square. It was difficult to get it to hold together being that there was no binding material. We thatched it on rafters of the same wood with the spiral branches before mentioned [rata] and used the spare boards for the floor. As we did not require a great deal of room, we arranged it for one bunk along each of the three sides while the chimney and door would occupy the fourth. We now built our bunks. Having plenty of skins we used one for the bottom and the others for the sides.

It took us several days but it was consoling to the mind, as well as for a useful purpose. I believe that I am within the mark when I say that I believe this was the only house ever built with only one nail – this being the only one we had. So taking things altogether, we were fairly well fixed.

[1] Turpentine tree or *dracophyllum longifolia*.
[2] Friday Island.
[3] Probably Auckland Island teal; size refers to juveniles, males and females. "The males have a beautiful green sheen on their head which really shows up in the sunlight." McClelland.
[4] Rose Island, which Holding refers to as Rabbit Island.
[5] Holding is in error here for the New Zealand fuchera is not found on the Aucklands.
[6] "thousands", according to Smith.
[7] Smith, *Castaways*: p.28.
[8] Musgrave, T., *Castaway on the Auckland Isles*: p.124.
[9] Captain Greig of the *Southland* describes it as having been "10' x 12', originally built of small branches and grass." (*Southland Gazette* 1865, p.120).
[10] Greig, J.B. *Southland Gazette*, 1865, p.120.
[11] Ocean Point is now known as Lindley Point. Baker, J.H., *A Surveyor in New Zealand, 1857-1896*: p.90.

18

Rose island

bearing N. ̂N.E.

"A dedicated philosophy of conservation is required to see and understand why these [islands] are not a place for risk-taking or compromise."

Department of Conservation Management Plan

THE next morning we awoke to a solid rain beating on the decks; however, we had not come so far to sit comfortably inside until it let up, which might be in minutes or in days. Weighing anchor, we headed towards Rose Island with the objective of finding the hut site. The gap between Friday Islet and Rose[1] was a sight to "try the stoutest hearts", for the tide was ripping through. Spume from waves crashing over the submerged rocks was flying high into the air. Holding would certainly have had to pick a calmer day than this to risk the crossing.

Rose is a relatively low (125 feet/38m) humpbacked little island, boasting very impressive cliffs on the north side. We slipped past the long fingers of wave-swept volcanic rock on which Holding's frail craft had finally come to rest, and turned in towards shore, where the DoC sign proclaimed: ROSE ISLAND – Entry by PERMIT ONLY. Whereas limited 'tourism' is allowed (under permit and accompanied by a Department of Conservation representative), at the Hardwicke site on the main island and on Enderby, landings on Rose Island are even more carefully controlled. Pete observed that many who want to get to the Aucklands make up a reason to do so, then try to get a permit, whereas we were granted the precious permit on the basis that the historical work we were doing could simply not be done anywhere else.

During the time of the Hardwicke settlement both Rose and Enderby islands had been used as pasture land. The cattle, which they left behind, were quickly killed off by sealers and Holding makes no mention of having seen any in 1865. A tiny note in 'Earth Watch', a syndicated column carried by the *Montreal Gazette*, reported in 1991:

A cattle herd left untended on a remote island in the South Pacific for 100 years is to be destroyed because it threatens the survival of a rare flightless duck, but an attempt will be made to save its genetic code. "This population has survived for nearly 100 years without any husbandry such as treatment for disease or parasites," said Hugh Blair of the Rare Breeds Conservation Society. They plan to take reproduction samples before the slaughter, and use them in research that may help modern cattle resist disease. The old strain of shorthorn cattle, nearly extinct itself, lives on sub-Antarctic Enderby Island where their predecessors were taken in an 1890 failed farming venture.

In fact, a female adult and a male calf were caught and returned to New Zealand for a breeding program. Those on Rose had died out earlier with the result that the regeneration of the tussock and scrub is further advanced than on Enderby. The rabbits placed on Rose in 1850 by Charles Enderby also played havoc with the natural vegetation. It is interesting to note that the rabbits on Enderby Island were not released until 1865 and were of a completely different breed, being Agente de Champagne or French Blues. Thus, to Holding, there was only one 'Rabbit Island'.

Their burrows created a severe hazard to sealion pups which would become stranded and suffocate. It has been estimated that the pup mortality sometimes could reach 10 percent in any one year.[2] In line with returning the islands to their natural state, eradications were "planned in close liaison with the New Zealand Rare Breeds Conservation Society and an alive capture operation undertaken."[3] Two aerial drops of poisoned bait in 1992-1993 achieved a 99 percent kill. The follow-up by dogs, traps and shooting resulted in the killing of the last 26 survivors on the two islands. All that remains today of a population once estimated as numbering from 5000 to 6000 are tiny bleached bones scattered over a now lush low ground cover. The next few years should see a rapid return to the natural high tussocks and the slow expansion of the rata forest.

In contrast to the main island, there is little vegetation on Rose. "The island is covered with hundreds of twisted, dead, dying, decaying rata trees laying about as if in a supernatural no-man's land. Yet, if one turns to the west, we see some rough scrub growth and then a whole area covered by tussock – almost as if it is a field which has been cleared by some farmer who, in the middle of the trees, planted some strange supernatural hay."[4]

It was surprising how quickly we had become adept at clambering over the rail, down the three-step ladder clamped to the side of the ship and into the bobbing dinghy. By the time we got ashore the sky had cleared to a brilliant blue and the heavy rain of the early morning had been swallowed by the ocean. We felt relatively confident of success as we spread out in a line 30 feet (9m) from the shore and made our way carefully forward, each studying the surface structure at our feet. The general location of the hut was clear from the manuscript and surely, even after 130 years, a building, which had had four-foot (1.2m) walls and fireplace, should not have disappeared entirely.

"There is no place here where they could have brought the rocks up from the beach."

"There are so many possibilities: there is a forested area with flat rocks nearby so he wouldn't have to carry them too far to make the fireplace, with a good view of the Port Ross Harbour..."

"My gut feeling is back there..."

"...reasonable shelter from the wind and fresh water nearby..."[5]

We covered the area, back and forth, finding several possible sites, but no rocks obviously piled by hand. It was most discouraging. It wasn't until our last evening on the Aucklands, while going through some papers which Pete had brought, that we found a reference from the Archaeological Expedition in 1972 to the hut site on Rose: "...all that remains of this 1865 site is a mound of flat stones about two feet high situated on a raised open flat area not far from the sea on the eastern side of the island. It is effectively disguised by a windblown prostrate *Myrsine divaricata*." The article went on to say that the site had been revisited in March 1986 and the location given was Sandy Bay, Enderby Island. I had seen this latter reference and knowing this could not be correct (as the *Invercauld* survivors had been rescued before they could move to Enderby), had not given the description of the site any great credence. Someone had typed 'Enderby' instead of 'Rose'.

The original map, made from Eden and Easton's survey during the war, has a site on Enderby marked as 'Invercauld Hut Site'. In his 1972 notes, Falla states that this is in error and correctly identifies this hut as having been built by the Southland Provincial Government or by a Captain Fairchild and was used by the *Derry Castle* survivors. The error in the location of the *Invercauld* hut site has been compounded by data by Fraser in *Beyond the Roaring Forties* (p.112), and the *New Zealand Geographic* article 'Wild Splendour', (Dec. 1990), both written after 1972.

Unfortunately, there was no map reference on the data we had on board, so on the last day, instead of trying again for the heights and possibly extending our stay, we returned to Rose. Armed with metal tent pegs we again lined up and probed our way around each myrsine bush. At one point I thought I saw a black rock in an unusual place. It was the flipper of a sleeping sealion just visible behind a tussock.

On our return to New Zealand, Pete looked up the original files and found a map reference to the site of the hut which turned out to be further inland and somewhat east of where we thought. The photograph which he sent shows the fireplace stones very clearly. It was unfortunate that we did not have this complete data before we left, for this site is the only permanent remainder from the *Invercauld* survivors. It is curious that while so many other historical sites are marked on island maps, this site was never correctly indicated.

After getting the camp in order my time was pretty well taken up with further explorations. Once in a while I got a rabbit (which were of almost all colours) by knocking it

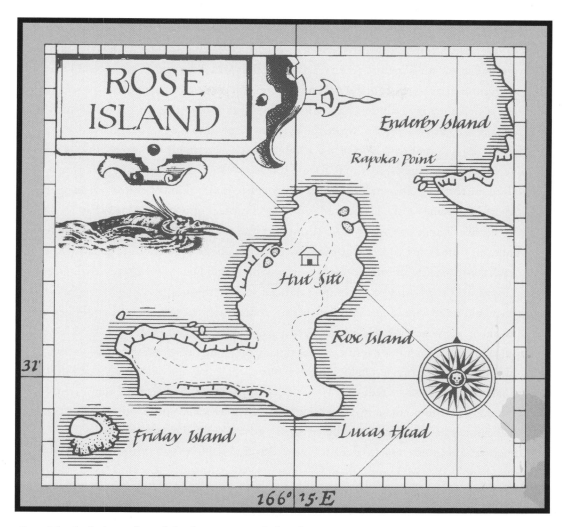

Rose Island, the last refuge of the three survivors before their rescue in May 1865.

over with sticks. Upon one occasion I had my adze on my shoulder and seeing one about fifteen yards distant, let go the adze and was surprised to see it topple over – but they were in very poor condition. Did I not wish I had a gun! I was always good at snaring them but here we had no wire and the skin of the seal was useless; especially as they did not make runs but ran all over the place.

There were some holes which I thought they used and hoping to find some, I began to break them open then to put my hand in. I was surprised to find something in there inclined to show fight, so opened it a little further and was not long in proving it to be a bird.[6] Anyone having been to sea will recognize them from the following description: they are black, except for a little white under the long wings and like all seabirds, have a hooked beak. They fly about in droves and dive the same when over a shoal of small fry. They then get up and off immediately. Well, can they fight when in a hole, as my hands would testify at that time, but I never left one until it was dead. We found them good eating but one peculiarity was that each one had inside it a bunch of green fat. I was at first afraid to eat that, as it looked so much like vergigrese,[7] [sic] but ultimately found it good. There is another peculiarity with them and that is that they lay only one egg, which is of a very large size for the bird which is in body, about the size of a pigeon, while the egg is as large as that of a duck's.

The seal did not appear to be cruising about as much now and it was rarely that we could get one, so we had to get something else or starve. Most of my time was spent wandering about looking for anything eatable. I now set to work to make a kind of crate with an opening. At one end I put some offal and lowered it in about four feet of water; the next day I took out twenty-six fish. There was one Mackerel amongst them. This was the only one we saw there.

About this time I used to go to the rocks by the shore on the seaward side, generally taking some sticks with me, to try to kill birds on the wing and did occasionally succeed. One day while I was strolling round by the cliffs, I noticed some birds on the top and thought what a treat it would be if I could get some of them. I found them quite tame and could get quite close to them, so went back to camp and got a piece of hide such as we laced the skins with. Scraping the hair from it I tied it to a stick about six feet long thus making a snare and was not long in catching six or seven. On looking over the cliffs, I saw one of the sights of a lifetime. There, about fifty or sixty feet down, was a ledge perhaps two feet wide and many yards long, running round a curve in the face of the cliffs which were in other respects quite perpendicular. The ledge was literally covered with birds in all stages of breeding.

Of course, I wanted some of them but how was it possible to get them? Yes, I must have some, so I will tell how it was done, because it beats conjuring. But first let me say how surprised and pleased the others were when I went down to tell them what I had seen. These birds are called Widgeon[8] and as large as a good-sized duck.

We had set apart one skin which had been stretched and dried for any purpose it might be needed. By this time I had become accustomed to cutting them into shoe lace size so I started on the job of cutting and scraping and was not long 'ere I had quite enough. This skin was rather stiff and would hold its position when once fixed, thus when made into a snare it would hang straight down. The wind at that distance would be an effectual barrier to my success. How could I overcome these two obstacles? Well, quite easily when you know how. Here is the trick. I first made a loop on the end, then formed the snare just above that. I doubled it, tied another short piece to it and on the end of this I tied a stone, then set my snare at right angles.

It was now a question of getting near enough to the cliffs to enable me to see the birds. This was a very dangerous position and in consequence I had to choose the spot with care and if possible, find a solid foothold. I was at that time prepared to take great risks, so was not long in obtaining a comparatively safe place. Fortunately there was little wind at the time. I dropped my snare over the edge and lowered it down carefully until it was just over the head of the first bird and let it go down with a jerk. This, as I had expected, tightened it round the neck of the bird without knocking it off and it was quickly on top of the cliffs and its neck wrung.

In this way I captured, in what I thought to be about an hour, twenty-six (including a few snared at the top). Sometimes they would disarrange my snare by pecking at it but that was all the trouble I had with the snare. This was pretty good work and I was well satisfied. I went to camp with fourteen, which was all I could carry at one time, then went back for the others.

Upon one occasion I knocked one down with a stone, or perhaps a stick, while on the wing. As it fell into the water a bird of the albatross type followed it down and the moment it was in the water began to devour it. There are several large birds of this class but not all called albatrosses.[9] They will follow a ship for weeks at a time, flying night and day.

While traversing the island on the open ground I had noticed several large birds, nearly all black with large body. I think they are called the Sea Hawk.[10] When I passed certain places they would almost hit my head with their wings.[11] There must be a cause for this, I thought, and soon found that it lay in the fact that they were breeding or watching their young. It did not take long to discover the young ones and I did get several despite the danger of getting blinded in their capture. I was able to knock over a few of the old ones, but found them very tough. I also got a few more of the small ducks, but none of the larger sized ones.

Now a word for the small birds. There were a few of what I thought was the Australian Robin and a green bird with pink eyes about the size of an English Sparrow.[12] They were so green that it was difficult to distinguish them in the trees. These were so very numerous that the place seemed alive with them and early in the morning they would begin to cherip. One day I saw two fighting and when they got on the ground I caught one of them.

We had often suspected another species, owing to having seen movements in the ferns; one day I caught sight of one in this way. I have not previously mentioned that the sealion always contained a lot of worms (I could with safety say quite a peck). Near where I had killed one I saw one of these birds feeding on the worms. It was long legged and long necked of a very nice brown colour with a beak like the partridge or pheasant.

Along the shore we had, as stated, seen some cabbage of wild growth in flower and some turnip in one or two places. I had decided to try to cultivate this, so set to work to make a garden, turning up the ground with the spade and fencing round the small plot with brush to protect it from the rabbits.

One morning after it was completed I saw one of these birds in there. So with the usual instinct for flesh, I went for it. It ran and put its head in the fence and I soon caught it. We had no immediate use for it as we had then plenty of food, so I thought how I would like to save it. If we ever got away I would take it with me. I set to work cutting sticks about three feet long and sticking them in the ground in a corner bounded by the chimney and the wall of the

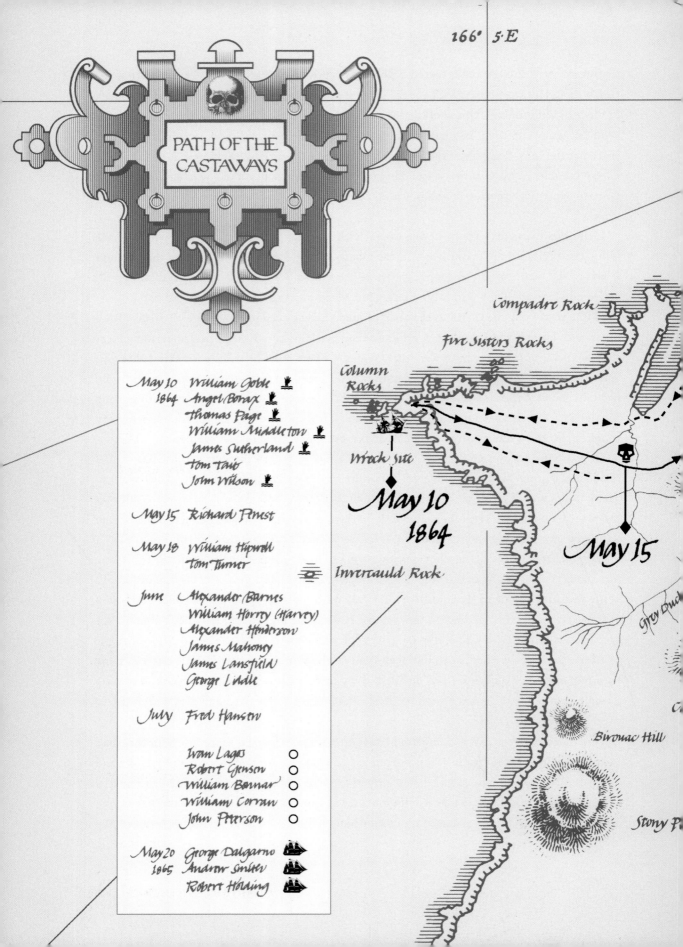

PATH OF THE
CASTAWAYS

166° 5'E

Compadre Rock

Five Sisters Rocks

Column Rocks

Wreck Site

May 10
1864

May 15

Invercauld Rock

Grey Duck

Bivouac Hill

Stony P

May 10 1864	William Goble	✋
	Angel Borax	✋
	Thomas Page	✋
	William Middleton	✋
	James Sutherland	✋
	Tom Tait	✋
	John Wilson	✋
May 15	Richard Penist	
May 18	William Hipwell	
	Tom Turner	☰
June	Alexander Barnes	
	William Horry (Harvey)	
	Alexander Henderson	
	James Mahoney	
	James Lansfield	
	George Liddle	
July	Fred Hansen	
	Ivan Lagos	○
	Robert Gensen	○
	William Bonnar	○
	William Corran	○
	John Peterson	○
May 20 1865	George Dalgarno	⛵
	Andrew Smith	⛵
	Robert Holding	⛵

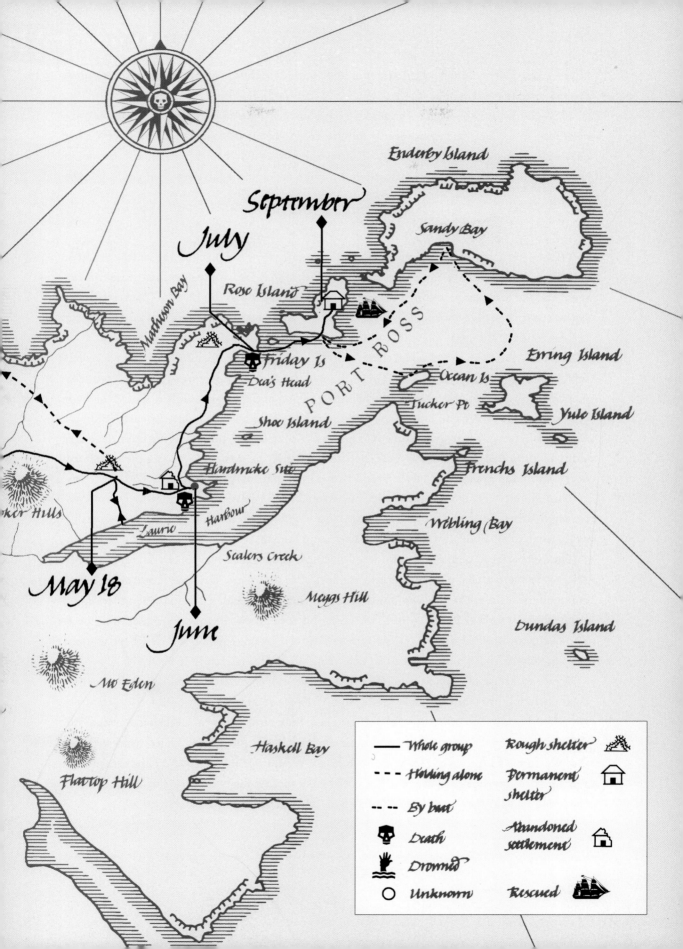

house, made a pen and felt satisfied it would be quite safe there. To my surprise and chagrin in the morning it was gone.

This was the only one we ever caught. I have since learned that in New Zealand they are called the kiwa [sic].[13] They are stated to have no wings but this one had one joint of a wing only and in consequence could not fly. There were some smaller birds which built in holes in the rocks, but they were insignificant and though we could hear them, we could not see them.

*   *   *   *

The roar of crashing breakers enveloped us long before we reached the northern cliffs. One thing I shall always remember about the Aucklands is the never-ending shriek of the wind and the constant pounding of the sea. Even from the top of the 120 foot (36m) cliffs, the sea swells still seemed enormous – even though we were assured that this was a calm day for the Aucklands. The spray blew into our faces and onto our camera lenses.

The cliffs ran in a great arc and there, below the edge of the precipice, was the rock shelf which Holding had described. I had never doubted it was there, but it was a thrill to recognise it. It was still early in the year for the breeding season and the vast number of birds had not yet arrived. However one mollymawk glided in and settled on the ledge. Robert Holding's 23-year-old great-great-grandson could not resist lying prone, head hanging over the edge, as his ancestor had done all those years before at exactly the same age. As a mother, I felt my heart was in my mouth, but I understood Bruce's compulsion to do this.

Throughout our stay we, too, became fascinated by the bird life. The pipits – brown, white-breasted little fellows about the size of a sparrow – ignored us completely. It was a fascinating experience to be among birds who have no fear of humans.

The yellow-eyed penguin is considered the most endangered species of penguin, for only 2400 to 3600 exist. They do not nest in rookeries, as do others, but hide alone in sheltered crannies under the rocks. We saw our first one nesting as we headed up a narrow waterlogged gully; slightly further Robin stopped abruptly, his foot just inches away from another nest. The occupant lifted her head, eyeing him calmly as if to say, "please watch where you put your great feet". Her mate was nearby, chest still dirty from his shift on the nest.

While making our way back to the cliffs on the extreme eastern end, we found two of these little fellows (they are only 18-24" tall or 46-60cm) standing motionless, 'wings' out-stretched, facing each other. As if choreographed, they turned their heads in unison, studying these strange creatures not dressed in their uniform black and white. There were so many that we found ourselves zig-zagging so that we would not interrupt their movements by getting between them and the sea. We could have spent hours watching the antics of these charming birds as they stared, waddled, stood, jumped and stared some more.

Here again, we found sealions far from the sea. Nestled under the protection of a large

ALL COLOUR PHOTOGRAPHS BY PADDY RYAN, EXCEPT P.188 (CENTRE LEFT); P.190 (TOP) AND P.192 (TOP LEFT): MADELENE ALLEN

(*Page 185, Top*) View from Dea's Head towards Friday and Rose Islands; (*p.185, Lower Left*) The sailors found mussels and seaweed a good source of protein and carbohydrate; (*p.185, Lower Right*) Basaltic cliffs of Dea's Head. (*This page Top Left*) Cape pigeon; (*Right*) Skua feeding on dead sealion; (*Lower Left*) Juvenile Auckland Island cormorant (shag); (*Lower Right*) Yellow-eyed penguin. (*Opposite page, Top*) Royal albatross is a spectacular sight airborne over the wild and windswept landscape of the sub-Antarctic Auckland Islands. (*Lower*) Giant petrel or 'nellie'.

The Auckland Islands are a botanist's paradise. (*Clockwise from Right*) *Bulbinella rossii* thrives on high rainfall of the islands; A relative of the carrot family, *Aristome antipoda*; *Epilobium confertifolium*; *Gentiana* sp; *Gentiana* sp; *Pleurophyllum speciosum*; Archegonisphores of a liverwort (*Marchantia*); Auckland Islands tomtit; *Stilbocarpa polaris*, one of the main food sources for the survivors, is rich in vitamin C.

(*This page, Right*) Hooker's sealion; (*Lower Left*) Auckland Islands shag (*Leucocarbo campbelli colensoi*) busy nest-building; (*Lower Right*) A Royal albatross breeding pair (*Diomedea epomophora*).

(*Opposite page*)(*Clockwise from Top Centre*) 1993 expedition crew member Eric Hayford-Cross on cliffs of Rose Island, with Enderby Island visible intersecting horizon; Red crowned parakeet (*Cyanoramphus novaeseelandiae novaeseelandine*); Auckland Island banded dotterel (*Charadrius bicinctus exillis*); Young female sealion sheltering from males amongst rata trees; Cow skeleton, Enderby Island.

(*Page 192*) Department of Conservation's Pete McClelland in discussion with 1993 Expedition members on Enderby Island clifftops; (*Inset*) Robert Holding continued working into his 80s; (*Top Left*) His final resting place at Chapleau, Ontario; (*Top Right*) The Southern Cross.

Robt Holding

tussock, a female raised her head to watch us pass, and went back to sleep. A great bull followed us through the tussocks and while we were concentrating on avoiding him, we heard a snort behind us and there was an even larger bull within a few paces.

It was here, on the high point of the island, that we saw our only 'finger post' which dated from the 1870s. Nine of these posts stand throughout the islands. These markers had been placed to point the way to the castaway depots which the New Zealand Government with the assistance of the Victorian, New South Wales and Tasmanian governments, had set up around the islands in 1868. After 1877 New Zealand took full responsibility for them and carried on the service until 1926.

In *The Castaways of Disappointment Island*, Eyre describes their arrival at the depot on the mainland: "We found the depot consisted of three sheds, all painted white. One of them was rigged up inside with bunks for sleeping purposes, one was the provision store, and the third – we could hardly contain ourselves for our delight – contained a wooden boat."[14]

What a difference one of these would have made to the *Invercauld's* crew.

Hiking back across the rugged volcanic rocks which were covered with white lichen, we marvelled at the variety of sea life and the colours. There were paua shells, mussels, and limpets in abundance. At the tide line, limpet eggs, looking for all the world like onion rings, clung to the undersides of sheltered rocks.

We were picked up at 4pm and motored back to harbour. What better thing to do after a long day's hike in the sunshine but go for a swim. Bruce, Pete and Paddy dived in from the side of the ship and surfaced very quickly. Attracted by the unusual splashings and cries, a couple of curious sealions came over to play, frolicking close to the side of the boat. The water was calm enough for us to see their sleek black bodies as they streaked just below the surface. A head would shoot up, whiskers twitching, to see if anyone was coming, while the other one would slide over, before rolling and spinning together. They paused again to look up at us with beautiful dark eyes, as if pleading for the swimmers to come back in. It was an incredible experience. I made the comment in my diary: "...amazing how less threatening they seem when looked down upon from an 80' boat than when they are telling us to get off *their* beach!"

That evening we had a memorable encounter with the local wildlife. Jim's tape describes it best as he was there from the beginning:

I was out on the stern deck with Lance. It was raining, and we were just standing there under the canopy looking down the bay. He thought he saw something and I looked and couldn't see anything, then all of a sudden a black object was visible maybe 300–400 yards down the bay from us as we looked inland. A short time later, I heard a noise, a kind of "whoosh". Just looked to our right and there is a whale lying alongside the boat – big – huge, in fact, and covered with barnacles. He was just lying there. There was a mad dash by everyone from the cabin out onto the deck into the pouring rain. He surfaced, played around the side of the boat, went under and then came up the other side. I think he was drawn to the sound of the generator. It was really something to experience!

It was magical to see this great sea creature, a southern right whale, lying on his side an arm's length from the boat. We could have stepped over the side onto his back; one great eye looked up, watching us all looking down. After our days of sightseeing, I had the feeling that when he returned to his pod he would tell them what *he* had seen in Laurie Harbour.

[1]   Friday Islet, 2 cables south-west of Rose Island, lies midway between Rose Island and Lindley Point.
[2]   Department of Lands and Survey, *Management Plan for the Auckland Islands, 1987*: p.19.
[3]   ibid: p.19.
[4]   Jim's tape.
[5]   Jim's tape.
[6]   Muttonbird, or sooty shearwater.
[7]   Verdigris is a "green or greenish-blue crystalline substance formed on copper, brass or bronze." (Webster)
[8]   Shag.
[9]   There are several species of mollymawk which are the largest in numbers of the albatross family. They are smaller than the 'true' albatross and the dark colour on their wings is continuous across the back.
[10] Skua.
[11] Sorensen, *Wildlife of the Subantarctic*, reported having been struck several times and told of a companion receiving a rather nasty scalp wound.
[12] Antipodes Island parakeet.
[13] This was likely an Auckland Island rail for kiwi are not known on the Aucklands. This flightless bird is extremely rare now; the Auckland Island rail is an endangered species only to be found on Adams and Disappointment Islands.
[14] Escott-Inman, H., *The Castaways of Disappointment Island*: p.252.

19

ocean island

bird of the island

Our supplies had become very much restricted, for the sealion became scarce and the most of the birds done breeding. We now frequently had to eat what had been hanging for many days and it was really disgusting to look at, so something must be done at once.

The mainland, about two and a half miles away,[1] appeared to extend about one mile further seaward than the place where we were. We could see breakers between it and the next island which might be near half a mile long,[2] then another island which looked almost to be connected with it.[3] We had often wished to see the bay and these two islands. We also had a good view of another island which was perhaps a little more than half a mile beyond ours in the seaward direction, which looked most attractive.[4]

One morning, when the water was smooth, the Captain and I decided we would venture out. We headed first for the inside island[5] and had a very nice trip, although very slow, owing to the construction of our outfit. It was a real pleasure to be able to get away from the island and be afloat again, even in our primitively constructed boat. I felt as if I had no further care and the trip seemed all too short. We found the island very rocky with no shelter and nothing of an enticing nature except roots, which were in abundance. There were no signs of seal or sealion. The other appeared to be of the same nature so we did not go on to it.[6] The Captain said there was a lot of mackerel, but as I was rowing I didn't see them.

When we were about half way across[7] we could perceive a sand beach of quite a size and as this was one thing we had been wishing to find, we became quite elevated, especially as we soon discerned some moving objects on the beach. We were puzzled for some time as to what they might be – at last we could make them out to be sealions. If they had been cannibals, I believe we would have chanced facing them.

The beach proved to be a very beautiful one which extended about two hundred yards along a kind of bay. There appeared very little wood near but some brush on the right hand side where we thought to land. The back of the beach was covered with the long flaggy grass. On the beach there were three or four sealions, apparently trying to sleep. Several gulls, some flying, others strutting about, were all trying to aggravate them; sometimes flying close to their heads and striking them with their wings. The sealions would bite at the gulls and wallow about; then begin to throw up their food which consisted of fish, as a matter of course. By this means the gulls got all they could eat.

Having been amused by them for some time, we rowed in to some rocks on our right which were about seven or eight feet high. In one place they had been worn down smooth where the seals had been climbing up and sliding down. This gave us a clue to their lair. We were not long in making the boat fast and started to climb up the rocks. I, as a matter of course, was in the lead, and had not gone more than ten yards when I spotted one laying asleep among the long grass. Turning around, I held up my hand to signal the Captain to stand still. I only had two or three steps to make to get a blow at him when I found he was awake. He popped up his head just in time to get one nicely on the nose and my knife was into him quick.

After making sure that he was quite safe we went through the low bush and into the long grass and soon saw another. This one met the same fate. We now had all we required for the time. Not wishing to disturb the others we quickly set to work to skin and cut up the ones we had killed and to take them down to the boat. We decided to take one in the boat and tow the other one behind. As we carried them over the sand beach we had to pass close to another one. During the time we were passing back and forth he watched us closely and came quite close. When we pushed off and got in, he took to the water and had quite a game around the boat.

I might here be allowed to say there is no comparison between these animals and what I have seen in captivity, as I quite believe some of them weigh quite six hundred pounds and when their skins are stretched many of them are six feet long.

Before taking leave of the sealions let me say there is an incident which I have overlooked. While tackling one about half-grown, it jumped at me over a log and ripped my pants twice. But I got him. This was the only one that ever shewed fight to me, for they are as a rule harmless, even though they always open their mouths, thereby looking ferocious. Now while we were there we killed forty-two and out of that I killed thirty-seven. So I think I can claim to have done my part.[8]

We got back with safety and with the two seals, so we considered ourselves well-paid for our trip. Having met with so much success and seeing where we could have lived for a long time, it was only natural that we wanted to go there. We discussed the matter and came to the conclusion that it would be worth while to move over there as early as possible.[9]

*   *   *   *

Enderby is a low-lying island, about two and a half miles long (4 km) and covering 1700 acres (688 hectares). Because of its beautiful beach and markedly sunnier climate it has been known as the Riviera of the sub-Antarctic. However, like the other islands of the Aucklands, rapid weather changes occur in this area and the *New Zealand Pilot* cautions:

Sandy Bay, on the southern side of the island, situated one and one half miles westward of its eastern extremity, is a convenient temporary anchorage, being protected from all except south-easterly winds, and the holding ground of tenacious clay.[10]

Before putting us ashore, Lance warned us that if the wind shifted we would have to move fast to get back to the boat because he wasn't going to wait for us. "It would be a wet, cold, windy, miserable night for those of you who get left. So if any of the crew tell you to 'git' then you 'git' in a hurry back here." Fortunately, the wind stayed steady and we were able to spend the full day walking the entire coast of the island.

Sixty-seven great bulls basked in the sun and only raised their heads in casual interest as we stood waiting for the second boatload to come out from the *Evohe*. The 'gentle' appearance of Enderby stands out in direct contrast to the rugged hills of the main island.

In the terrible days of the Hardwicke settlement, the inhabitants enjoyed picnics and were rotated for short stays to enjoy the beach and the sunlight during the summer months.

I had the pleasure of accompanying Mrs. Munce, Mrs. Barton and the families and also Captain Glennys and some of the Fantome's officers, to Sandy Bay where a constitutional run on the sand, as well as dinner was indulged in – and all returned in good spirits before dark.[11]

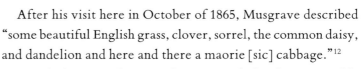

After his visit here in October of 1865, Musgrave described "some beautiful English grass, clover, sorrel, the common daisy, and dandelion and here and there a maorie [sic] cabbage."[12]

With their mandate to release animals for future possible shipwrecked mariners they had left 10 goats[13] and 12 rabbits on the island. Over the years, the animals, plus the firings to keep down the rata and the tussock grass to improve the grazing, caused inestimable damage to the vegetation. "Large areas of Enderby Island resemble a war-scarred landscape, with lifeless trunks and barren branches standing or lying uprooted on the sward."[14]

We sauntered along the beach, wondering at the tremendous contrast of this magnificent crescent-shaped beach compared with the black volcanic rocks of other shorelines and the sweep of low green grass with the intertwined vegetation which had greeted us on other landings. Before long, however, the beach gave way to rock with a shoulder-high ledge about 10 yards (9m) back from the shoreline. It would be this portion to which the largest beach master had laid claim. We spread out, and in single file under his watchful eye, to the accompaniment of many snorts of warning, crept past. Once past the cordon of bulls we cut across the sand dunes and through the myrsine and tussock towards the cliffs on the north coast.

Enderby is one of two nesting sites (with approximately 60 breeding pairs) of the royal albatross. The other, larger site, is found on Adams Island which has approximately 4500 breeding pairs. On the voyage down they had been constant companions and we had been enthralled, but here we were part of their land world. They showed no fear or alarm, whether on the nest, in the air, or while performing a mating dance; only a moment's pause to look at these two-legged creatures without wings or flippers. It was a very humbling experience to encounter them and to see a 'royal' in flight over land.

The albatross must be one of the most beautiful birds in the world – impressive by its sheer size, elegant with its graceful soaring, magnificent in the purity of its plumage, and just a little intimidating with the hook on its beak. Yet, standing just at the border of our five meters, watching a female on the nest, the bird's great eyes seemed almost endearing and gentle.

I regret that I am not more knowledgeable about birds for the Aucklands are an ornithologist's paradise. All I was able to do was admire their beauty and marvel at their tameness. Walking across Enderby, I would never have guessed that the little banded dotterel is an endangered species; the 1987 DoC report estimates their numbers between 100 and 200. Two of the rare red-crowned parakeets were sitting waiting for us when we landed on Rose. We saw petrels and prions, which are described as scarce, though I would hesitate to guess which of the 10 species they were.

We had a wonderful half-hour with four Auckland Island shags at the top of the cliffs. These birds are about the size of a duck, but much finer featured, with dark greenish-blue backs and satiny-white chests and stomachs finished off by a little stubby tail which balances them as they totter about on large webbed feet. They are an endemic species and have been placed by the Wildlife Service in the 'rare' category. Spotting one of the 5000 waddling along the cliff, we seized the opportunity to sit down and rest and observe.

It was soon joined by its mate and they began collecting material for their nest, waddling back and forth within five or six feet (2m) of us. Possibly feeling left out, another landed about two paces in front of Paddy Ryan's camera lens.

The southern skua, or sea hawk, is the scavenger of the islands. This dark brown bird with white wing flashes is about the size of a large duck, but with a vicious looking hooked beak. We had our first close-up view of two of them having a lovely feed on a very dead seal. Standing no more than 15 feet (5m) away, we watched in fascination. They simply turned their heads, asking if we would like to join them, and went back to their feed. Half an hour later, while we were enjoying our own food, one came over to see what we were offering. It landed right by my foot and proceeded to check out the possibility of any seeds which might have lodged in the tread of my boots. Soon disappointed, it began to explore the hook of my gaiters, then onto the lacing and without hurry worked its way up the creases of the gaiters. When Pete commented "Mmm, a nice piece of Canadian thigh", I thought it time to end our friend's exploration. Slowly, so as not to frighten it, I removed the video camera from my eye and got up. It simply cocked its head, somewhat insulted that I had moved, squawked gently to inform me that I was far too alive to be really interesting, and lazily took off.

As we made our way to the cliffs we saw a plant with a beautiful yellow flower. Imagine our surprise when coming over the next low ridge we found acres of *bulbinella rossii*[15] spread before us. We had taken great efforts over photographs of the lone plant we had seen earlier. (It was at this point that I realized what a wonderful guide and mentor we had in Pete McClelland of DoC. He allowed us to discover – he could so easily have said, "don't waste your time here, there is a field of them on the other side.") The brilliant yellow flower, much the shape and size of a 'cattail', grows on a single stalk nestled in deep green lily-of-the-valley type leaves. It was impossible to walk without treading on them. Indeed we had to take care wherever we put our feet on this island. Jim's tape brings back memories of that wonderful day:

We looked down at a horseshoe-type bay. The first two-thirds of the horseshoe is hard, volcanic rock which then merges into a beautiful white beach at the far end before narrowing to a point. Past that is a bit of a bluff, going up gradually to about a 150 feet (45m). There must be 15 sealions sitting on the beach. It is a beautiful spot. There are penguins up above the rocks... just coming up the grassy area above the rocks. Then the land goes back into rolling hillside again and on to the thick rata trees.

We were sobered by the sight of *Derry Castle* reef. On a wild March night in 1887, the *Derry Castle*, only 100 yards (90m) from safety, smashed onto a low shelf of rocks hidden beneath the surf. Fifteen people were swept to their deaths. The eight survivors buried the bodies which they were able to recover. The figurehead is now in the Canterbury Museum in Christchurch. Selfish souvenir hunters have removed all plaques and the only marker today is a single upright stake and an iron bar from the wreck, lying half-buried in the grass. The punt built from the wreckage to get to the castaway depot in Erebus Cove had only recently been shipped to New Zealand for display in the sub-Antarctic exhibit in Invercargill's Southland Museum.

It was near to here that we had our most interesting encounter with a very irate bull sea-lion. We had been keeping an eye on this chap as he made his way purposefully towards us from several hundred yards away. This was the only one which had actually seemed to come after us. Paddy and Eric had been delayed taking photographs and had only just sat down for lunch. The rest of us were individually ambling around the area. Suddenly we heard a yell, turned around, and there were Paddy and Eric eyeball to eyeball with the bull.

I would have retreated quickly, but Paddy said he'd been there first and was going to finish his lunch. The bull feinted and lunged, retreated, roared, advanced, and generally made a nuisance of himself under Pete's watchful eye. Finally deciding that they were not a threat to his lady he moved off, casting glances over his shoulder with great swayings of his head.

Hooker's sealions are "the rarest and currently most endangered of the five species of sealion in the world."[16] The Auckland Islands are the centre of this world population which is estimated to be between 5500 and 6500. It is estimated that the rookery on Enderby numbers about 1200. When we landed at the beginning of December, the bulls had arrived and the females were beginning to appear. While we did not see many females on Sandy Beach, the coast at the extreme eastern end of the island appeared from a distance to be covered with rocks which, as we got closer, took shape into hundreds of beige females guarded by dozens of bulls. The harems were forming, but the mating does not take place until the end of December after the pups are born, although we passed by one amorous couple who ignored us completely.

It was with difficulty that we tore ourselves away from the colony. We could have spent the whole day watching the young bulls sparing, the old bulls checking their harems and the gulls swirling and darting. The shadows were lengthening as we made our way back to the sandy beach.

It was along the southern coast that we saw our first 'nelly' or southern giant petrel chicks. These large balls of fluff looked quite friendly but Pete warned us not to get too near, for they have the disconcerting habit of projectile vomiting when alarmed, hence the nicknames 'stinkers' or 'stinkpots'.

At one's approach the sitting bird utters a rasping note which terminates in a coughing squawk. This is the warning to keep away. Should one continue to approach, the bird will almost invariably eject a quantity of evil smelling oil with skill and precision. The chicks do exactly the same, and I well recall the job I had to clean my camera lens after trying to photograph a chick at the distance of fourteen inches; the chick's aim was perfect! They gorge themselves on carrion until often they are unable to take wing again. They then sit around on the ground, or fruitlessly try to wash the oil and filth off their soiled plumage in the water. If disturbed whilst in this gorged state, they immediately vomit the stomach contents until they are light enough again to take off.[17]

This chick would grow to have a wingspan of about seven feet (2m). The adult bird is a fantastic flyer, skimming low along the water and then, like a fighter plane shooting upwards, swirling and diving straight back down.

We were suffering from information overload; I had difficulty in distinguishing between the nelly and the white-capped mollymawk and between the royal and wandering albatross, even though Pete, patience personified, identified them over and over for us.

The beauty of the wildlife and windswept landscape was overwhelming. Sealions lying in stream beds, two female sealions having a 'natter' in the sun, bones of cattle bleached by the sun, fantastic shapes of the rata trees, flocks of white terns swirling and crying, two penguins staring at us as we pass by, tiny inland ponds reflecting the sky. The sun had disappeared behind the heights of Mt Eden by the time we returned to Sandy Beach. After another day of being surrounded by bull sealions on the island, we marched past the 67 without even thinking about them, even the one we had to squeeze by to get to our boat.

The *Evhoe's* anchor chain was pulled aboard and we turned our backs on sunlight and faced the mainland, where the sky was overcast and we sailed into the rain. I could almost feel the pages of history turning back, revealing longboats carrying refreshed families back to the settlement at the end of a joyous day at the beach.

[1] Tucker Point.
[2] Ocean Island.
[3] Ewing Island.
[4] Enderby Island.
[5] Ocean Island.
[6] Ewing Island.
[7] To Enderby Island.
[8] Some accounts imply that they caught very few.
[9] To Enderby Island.
[10] *New Zealand Pilot*, 12th edition, (1958): p.446.
[11] Macworth, Wm., unpublished diary. Quoted from *Beyond the Roaring Forties*: p.106.
[12] *Transactions and proceedings of the New Zealand Institute*, Vol. XXII, 1890. p.82.
[13] In 1987, 58 goats were captured and taken to New Zealand for research and husbandry. 105 were shot in 1989 and they are now believed to be totally eradicated.
[14] Fraser, C., *Beyond the Roaring Forties*: p.137.
[15] Of the lily family.
[16] Department of Conservation, *New Zealand's Subantarctic Islands*: p.51.
[17] Sorensen, J.H., *Wild Life in the Subantarctic*: p.23.

20

Saved a:

cast ancho

We had now meat for a long time if we could only have kept it, but the flies soon spoiled it for us. We knew from what we had seen on this trip that we could get more any time we could get across, so our minds were at ease on that score, but there was more to be done.

There was, as previously stated, only one kind of wood on that island that would float and as there was little that would be of use to us at the beach it was necessary to take some across. With this in view, I set out one morning with the old adze, the only tool at my disposal – to get some of the largest sticks of the soft wood. After getting a nice lot went back to the hut. When about twenty yards away, I saw to my horror that the roof was on fire. I was only just in time to save it by beating out the flames. The mate had been burning something more than usual and had set it on fire. There was more trouble for the mate.

In this way two or three days passed and one day in getting wood to take over to build a camp. I had found a few sticks and cut them ready. After being out perhaps an hour and a half the handle of my adze broke, making it necessary to go back to camp and put in a new one. This was never done, for nearing the hut I heard such an unusual noise that I became quite alarmed and started to run as fast as my legs would carry me thinking that our camp was again on fire. On arriving I found that the noise was coming from the direction of the bay. When about half way there I could see the Captain was calling and waving like a madman. I tried to calm him down so that I could understand him and asked him what was the matter. The only thing I could make out was... "A SHIP A SHIP A SHIP A SHIP."

By this time Smith had returned and the Captain upbraided him for being away, saying that he had been calling for an half hour. It was with difficulty that we could quieten him and get something like definite replies to our inquiries. Having got him calmed down, he said he had been looking across the bay when he had seen a full rigged ship pass to the southeastward outside the breakers, between the Main Island and the one that he and I had been on a few days before (on the other side of the bay). She was under reef topsails and making for the entrance to the bay and that she was then cut off by the islands and that we would soon see her coming round the point. He said, "if you look you will see it pass between the two islands", but we soon found that the land was too high for that and the breakers extended out for some distance so therefore she would have to stay away out.

We were left in suspense for what seemed a long time. At last we did see her, I should imagine about three miles out from the point, which would account for the length of time she was invisible to us. The main thing was to attract their attention. The only thing we had in the way of a flag was a blue shirt. We were on low land and, in the absence of a flag staff, must make plenty of smoke. I told them to make a good fire with seal fat and green bushes while I would run up to the top of the island and put up the shirt on a stick as near the skyline as I could. This did not take me long. They had now a good fire and soon had lots of green branches on it.

We then sat down with tremulous hearts for although the ship was approaching the centre of the bay but she appeared to be going right away. Just as we were giving up all hopes, she went around. Smith declared they had fired a gun and in a short time he said they had lowered a boat. He then said that they had fired a gun in the boat. Neither the Captain or myself had seen these things and we were almost incredulous on the matter. The ship, however, was making across the bay and they fired another gun in the boat, which we did see. This, of course, gave us all fresh life and we fired up in consequence. The boat was now plainly to be seen rowing up the other side of the bay.

We had now the consolation that if by any means they missed seeing our smoke we would be able to catch them somewhere with our punt. At last we saw that they had seen the smoke and were making straight towards us. It would be difficult for me to describe my feelings as they drew nearer and nearer and I don't think there was a dozen words spoke all the time they were rowing across.

But when they were very near, the Captain said, "Don't speak to them, I will speak to them". This was needless, as far as I was concerned, for I could not speak for some little time. They pulled in close to where our people had capsized and he directed them to go a little further on so that we could pull the boat out.

I have entirely forgotten what was the first greeting, but I do well remember that the man in charge of the boat said, after he had heard our position described in a few words, "Well, my poor friends, our ship is leaking, but I promise you, I will take you with us."

After getting the boat secured, by pulling her well above the high water line, we began to exchange tales at once. Our position partly explained itself, for having been there over twelve months, amongst dirt, grease and smoke without the necessaries to enable us to get a good wash and with our clothes covered with grease, we were in a pitiable plight.

We now learned that the boat's crew consisted of the Boatswain, three white men and one black, all speaking good English. They in turn stated that they belonged to a vessel called the *Julian*, under Portuguese colours; that they were from Macao with Chinese to Callao.[1] Their ship was leaky and they thought there was a naval rendezvous on the island. I hope to explain this later. All they appeared to have brought with them was some bread and a musket with some powder. We, of course, took them up to our mansion and soon had some meat cooked. It was now dark and the only light that we had was an oil lamp, which I had made by cutting a glass bottle and putting a floating wick in it. As may be imagined it gave more smoke than light, but was better than none. Although we had been without bread so long I had no craving for it at that time.

We, of course, had quite a long chat after supper, but I noticed a certain amount of reticence on the part of the boat's crew on some subjects. They had not been there long, when I found a disagreeable stranger on me and learned in the morning that the ship was smothered with them. This was disagreeable news, but there was no choice in the matter. We managed to make them fairly comfortable inside and I think they had a good sleep. We were early on the move to get off to the ship. While they got breakfast and took out the skins, which the boatswain wanted to take for chafing gear, I took the musket and some small stones in lieu of shot and went for a rabbit. That did not take long and I came back with three. I then had a bite and we loaded up the skins, nine in all. We then took our places in the boat and left at seven o'clock by the Boatswain's watch.

---

[1]   Baker identifies the rescue ship as an American whaler which was passing the island. Dalgarno and Smith identify it as a "Spanish Brig"; the *Aberdeen Journal* account says a "Peruvian vessel".

21

the ship

miles distant

"These islands are amongst the highest-valued conservation assets in New Zealand and the world, and have to be managed on this basis, in perpetuity."

Department of Conservation Management Plan

WE had spent only seven days on the islands. All too short a time; but, then again, we had a boat at our backs, good food, and a choice of when to leave. On that last morning, I stood on the deck of the *Evohe*, and in my mind's eye pictured the longboat pulling away from the *Julian*. This harbour, so empty, so peaceful, had once held great sailing ships – and that one in particular which had brought long-awaited relief. A sobering thought when I considered that it was that particular ship which had resulted in my brother and me (and our two sisters and brother at home, and of course our mother) *being* here at all.

Our estimated time of departure was 5pm, which would bring us to the southern end of Stewart Island the evening of the next day. Only a few hours left, either to visit Ocean Island, or try again to find the hut site on Rose Island. We settled for the latter, as the urge to identify, to actually touch something that our forebear had touched was too strong to be denied.

We spent a frustrating morning searching again for the fireplace and gave up our quest about 1pm. Brenda had a sense of satisfaction that day though, as she had spent an exciting half-hour sketching two Auckland Island falcons which posed motionless beside her. Paddy had wanted to do some underwater photography, so while the rest of us finished our notes and prepared for sea, he and Lance went for a quick, cold dive from the side of the boat. At 4:30, a dirty black sky before us, and a backdrop of hills shrouded in mist, we weighed anchor. It was over. With a sense of sadness, exhilaration and accomplishment we motored slowly past Rose and Enderby for the last time. Sails set (unfortunately splitting the mizzen in a wild gust), we headed off into seas that seemed distinctly higher than when we had come down.

My diary, written in a somewhat shaky hand that night, on the sea once again: "10:45, 9th December: This is really a rough trip, but have absolutely no fear. Sounds of the sea... whistle of the radio, crash of things falling in the galley – glad I don't have to clean it up! The boat is pitching, and a great THUMP as it comes down after a particularly high swell. The waves are four to five meters high and our course being almost due north (352°) is putting us right in the troughs. Horizon frequently disappeared through the saloon portholes. We are running completely under sail, which seems appropriate."

Our only slight mishap and my only moment of real fear came in the middle of that night when we broached. I woke up to the most almighty crash and the sense of the boat slewing violently, then heeling sharply, throwing me onto the stormsheet, the porthole above me. Before shock could turn into terror, the *Evohe* righted herself and sleep came again surprisingly quickly. My husband slept undisturbed.

I had fun with my video camera the next afternoon, trying to get footage which would convey to my landlubber friends what the movement of a sailboat in the Southern Ocean was really like. The most successful shot was taken of a tap with the water swinging left and right of perpendicular by about 45°.

We reached Stewart Island at 11pm the next night, 31 hours after weighing anchor. Under power we slid into the magnificent harbour at Pegasus – hills black against the still grey sky – and dropped anchor. The stillness was uncanny and the sea around us was like glass.

There is always something bitter-sweet about the last night of any special endeavour, especially when new friends live so far away. Well bundled-up, Jim, Pete and I sat on the stern deck far into the night finishing off our drinks while the rest slept below. The silence was perfect except for the hoot of moreporks (New Zealand native owls) and the greatest thrill of all – the cry of a kiwi. We spoke of "seas and ships and sailing wax, of cabbages and kings." My only regret as I went to bed was that the sky was overcast. We had been eight weeks in the southern hemisphere and I still hadn't seen the Southern Cross.

"Are you awake?" It was Jim whispering into my ear. I had no idea how long I had been asleep. "The Southern Cross is out." Throwing on a sweater, I shot back out, and there it was, swinging high: five faint tiny stars briefly glimpsed between the clouds.

The next day, even though Mickey took us on the 'inland passage' of Stewart Island, an exhilarating, heart-stopping sail, the let-down set in. No longer were our hearts yearning for new adventures. For our crew members the return to families was at hand, and for us, the long drive back to Auckland and the flight home. Seven hours from Pegasus to Bluff – the first sight of the mainland, the lighthouse, the harbour buoys, and we were beside the quay. As Eric tied the hawser to the bollard and the engine shut down, I felt a sense of sadness that the adventure was over, but knew that our experience would have a lasting effect on all of us.

The Captain's Gig was a good one with the usual four long ship's oars. The boatswain was in charge with our captain beside him; Smith and myself doubled up on the two oars. I might say that there was a pretty fair load with the nine skins in the bottom and eight men. We had taken the precaution to take along the old water can to act as a bailer in case of emergency. This is the one which I had put a wooden bottom in. It did ultimately save our lives. After scanning the sea front, which we took to be fully three miles wide, we could see nothing of the ship. We rowed straight across to the island across the bay which was fairly smooth. When we got to the island [Ocean Island] the boatswain and myself landed to look for the ship. The island being small, we climbed to the top, but could see nothing of it. The thought of us all was "What if she had left!"

The sea was running high outside but the boatswain decided to go out to try to find her. We then had perhaps two miles to go to catch the rough water and the prospect looked very gloomy. However, before we got out to it, we caught a sight of the ship, only just visible and going from us with reef topsails! We were not yet into the rough water but steering across the wind and the boat began to ship a lot of water. Once we lost sight of the ship for a long time. Needless to say we were all very downhearted.

The Captain had taken the Mate's place at the oar and we were all about exhausted. The mate had been appointed to bail out the boat, but seemed too far gone to care. Knowing the danger we were all in and the kind of man he was, I took the matter in hand and he was soon bailing again. It certainly looked as if we were doomed as we were nearly out of sight of the land, and to think of going back in that gale was out of the question. When nearly all hope was gone, we fortunately caught another glimpse of the ship. The fact that she was still going away from us was decided by the long time we were pulling after her without any apparent effect. I shall never forget the state of mind we were all in. We could not now see the land and only with difficulty could the mate keep the water down. To encourage us the Boatswain kept telling us that the ship had gone around. But when we turned to look we knew differently. I am not exaggerating when I say there was not one in that boat who was not crying except myself.

I urged the boatswain to throw over the skins. He refused. At last I told him it was useless and discouraging to continue telling us that the ship was getting nearer and disappointing when we found the contrary was the fact. We then rowed in silence for some time. Then at last he said, "Well now Bob, you look. She is round."

I did look and was surprised at the difference in her appearance. No one who has not been in a similar position could believe the great change in the men. All worked like mad. We could almost momentarily see how fast we were approaching each other. Almost before we knew it we were near the bows on the starboard side. The powder had got wet so we could not fire the gun. As we passed round the bows the Boatswain said, "Don't call out, I will do that." He did, but no one heard him and the ship was going by when I said, "All call out together." This had the desired effect and both the helmsman and the captain, who was on the poop, heard us. The Mainyard was backed and we were alongside in a giff.

Officers first being in order, they were soon on deck and the others were equally anxious. I told one to stay in the stern and I would hook on the bow tackle. This worked admirably and having lots of Chinese to help we went up with a Whoop.

To say there was a commotion on board would be putting it mildly. Everyone was surprised at our arrival and especially so on account of our deplorable plight. None had trousers below the knee and what we did have was saturated with the oil from the sealions or dirt or in combination. It took but a short time for an understanding of our position. The officers were taken into the cabin and me into the house on deck with the Carpenter and others. I was handed a stiff glass of Rum and put to bed at about twelve-thirty.

Being pretty well exhausted I soon fell into a sound sleep but that did not last long for I soon became fidgety, being annoyed with the crawlers. I can safely say if there was one in my locality at any time I was sure to get it. I soon learned that everyone on board was effected [sic] with them, but I may have got more than my share, owing to everyone being anxious to do what they could for me, packing the blankets around to warm me up. The crew was a mixture of nearly all nations. They were, however, all very kind to me and soon furnished me with an outfit consisting of a complete change of clothes. I never knew what became of the old ones.

It appeared customary on board this ship, in the cabin at least, to have an early afternoon tea, so about 5 p.m. I was told to go into the cabin for tea. I found all the officers present, including the Boatswain, together with Dalgarno and Smith. They also had got a change of clothing and looked much improved in appearance. Our meal consisted of eggs fried with sliced and fried potatoes, bread, butter and a few little delicacies with a glass of wine. The latter is accounted for by the ship being Spanish with officers of the same nationality.

I was not long in learning the reason for her flying the Portuguese colours. It was this; they could not carry on this trade of dealing in coolies otherwise, the slave trade being illegal under the Spanish flag.

The Boatswain belonged to Jersey and could speak French and Spanish as well as English. He was therefore a go-between for all purposes. The carpenter also had his meals in the Cabin. The Steward was French and I don't know if he spoke any other language. All this may appear foreign to the matter relating to my tale but the significance will appear as I proceed.

We lived very well on board, having coffee twice before breakfast and four meals per day. Owing to the mixed nationalities there was little conversation at the table. I soon began to learn that all was not sunshine on board. They had left Macao with 350 Chinese and about 50 had died, including the Chinese doctor and the interpreter. There was, in consequence, various sinister rumours as to the cause of some of the deaths which may be summed up in the following manner....

On board this ship the Boatswain was supposed to class next the captain. While I was on board he had charge of the medicine chest. The Boatswain and the Steward were supposed to be strict friends and each of them was very conservative in their conversation. It was stated that they had made the agreement that if either died on the voyage the survivor was to have the property of the other.

It now came out that the Boatswain's watch had been borrowed by the Steward, together with a certain amount of money that we owed to our relief. It was the Boatswain who had mistaken the Auckland Islands for Auckland, New Zealand and who had told the Captain that there was a naval station there. It was also frequently stated that it was his intention to desert at Auckland. They say it's an ill wind that blows no one good and this may apply in our case.

It was also common talk that the interpreter and the doctor had been poisoned and that the Boatswain had received $5.00 for sitting on the inquest of each of them. There may or may not be some truth in these rumours, but the reader may judge of that by what I shall state later on which actually came under my notice.

The time seemed to hang heavily on my hands and I offered to go to assist in sail making, and mentioning it to the Boatswain he assented. Having some occupation kept me from thinking of other things. I continued to work at this for several days and felt more easy in mind. One day I had been speaking to the man working and the Boatswain came and told me I must not speak to him. I dropped the sail there and then and went to my berth, determined to do no more. I had been working voluntarily and getting nothing for it and could see no harm in speaking. He always appeared to treat me with indifference after, and, I may say, that after the rumours about I always kept my eyes open for trickery.

We were getting well on our way to Callao when the Steward was taken sick and was strictly under the observation of the Boatswain. Two days later he died. I might here say that they occupied the same state room and that at the table, I sat directly facing it. After a burial service by the Captain he was cast overboard. I think I am justified in saying without mourning, owing to his unpopularity.

Burials were quite frequent all along, but they were Chinese who mattered little. Anyone could be walking amongst them while in their bunks and see them picking vermin from their heads and perhaps half an hour later see them dragged up the ladder and thrown overboard.

After passing the islands of Juan Fernandes, we had a little excitement by the capture of a porpoise, in which I took part. The Boatswain's mate had been whaling a good deal and as soon as the word went round that there were porpoises over the bows he was out with his harpoon. Having often talked with him on the subject, I was convinced that we would have some fun.

I might here explain that to ensure success it is necessary to have what is called a tail block fastened on to one of the forestays with a rope rove through it and a running boling on the end to throw over the tail when he is struck with the harpoon. Quick action is necessary, as the

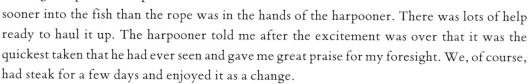

draught of the ship passing through the water is apt to withdraw the harpoon and the same applies to hauling it up. The harpoon was no sooner into the fish than the rope was in the hands of the harpooner. There was lots of help ready to haul it up. The harpooner told me after the excitement was over that it was the quickest taken that he had ever seen and gave me great praise for my foresight. We, of course, had steak for a few days and enjoyed it as a change.

About this time the Carpenter was taken sick and we naturally expected him to go off as the others had done, but after about three or four days he began to eat a little but he continued very weak until our arrival in Callao.

Just before we arrived in Callao we saw the largest school of porpoises I had ever seen. They reached as far as the eye could see, on both bows and the same in front of us. It was a sight to remember.

We dropped anchor on June the 26th and the mail boat which arrived that morning brought the news of the assassination of President Lincoln. Two days before we reached Callao the Boatswain took sick and requested to be sent to the hospital. He received permission to go on shore with Captain Dalgarno and Smith, even though the ship was quarantined. He was preparing to leave while I was at dinner and, as I have stated, I was sitting right opposite to his berth and the door was open. He was on his knees in his bunk and I saw him put two heavy bags (which looked like stockings) into the breast of his shirt. It was common talk that the Steward had had goods on board to the value of over $400.00 while the Boatswain had nothing.

I have not mentioned that there was also a passenger on board, Spanish by birth, who I believe, had cargo on board. He had, of course, come from Macao where he had been in business and I have reason to believe that he knew something of the Boatswain's doings there. It was what I learned from him, more than any other source, that I learned to distrust the Boatswain. If half of what I heard was correct, the latter was not to be trusted in any way. There is one thing certain, and that is that his sickness was of a premature nature, as the next day he came back on board and fetched a double barreled muzzle loading gun which had belonged to the Steward. He was never at the hospital at all. I went ashore with him and he asked me to carry the barrels while he carried the stock. He took me to a store on the mole and the keeper of that store went to the back with him. While they were there I took the precaution to sound the barrels, but found them empty. While we were in there, the storekeeper got me on one side for an instant and began to question me in relation to the Steward's death, but the Boatswain did not allow our conversation to last long.

All the foregoing helped to convince me that everyone was suspicious of him. The passenger went so far as to tell me that he always carried a long crooked bladed knife for protection and that he would not trust the Boatswain anywhere. While some of the others who had been in his company in Macao, told me that he had been in the habit of knocking Chinamen down and robbing them. Perhaps it will be thought that after the service he had rendered to us that I ought not to mention these things. I may therefore excuse myself by saying, it is with no vindictive spirit that I do so, but with the hope that it may be of service to some of the younger element as to their choice of strangers as friends at all times.

* * * *

The story that Holding tells of the "temporary" disappearance of the *Julian* after dropping their whaleboat, leads to all sorts of speculation in light of his revelations about the Boatswain. Certainly, it might be argued that if there had been an east wind, or a strong westerly, the waters near the mouth of Port Ross Harbour would not be as safe. The *New Zealand Pilot* states "...at the entrance to Port Ross squalls of great violence blow down from the hills during strong westerly winds." However, Holding mentions that it was relatively calm. Why then, did the *Julian* sail away? There are many accounts in seafaring literature of the time of particularly disruptive or untrustworthy crew members being abandoned on convenient islands. Did the officers have this in mind for the Boatswain, regardless of the others with him, and plan on leaving shipwrecked mariners to their fate? We will never know.

There had been an hurried subscription taken up in our behalf, and the Captain's fare to England was secured. A storekeeper told me that he held $40.00 for me which I was to receive on leaving.

Smith had gone to the hospital in Callao. I met our Captain who took me to the British Consulate and introduced me to the consul. He had little time to spare as he was leaving on the mail boat that day. Having been taken into the Consul's office, where I met the Commodore, I was there subjected to a long questioning as to our sufferings and ordered to a boarding house. The Commodore inquired what I wished to do, and said would I like to join H.M.S.[1] If so, they would do what they could for me. I thanked him and told him that after what I had gone through I wished to go to England to see my parents. I then asked for a certificate, showing that I was one of the survivors. This I obtained, and it is still in my possession, though much dilapidated. Again thanking them and bidding them Good day, was taken in hand by a boarding housekeeper.

Some of the crew of the *Julian* had promised to give me some money as they were being paid off that day, so I saw three of them and they gave me $5.00 each and I was very grateful for it. They had been very good to me all through, and in fact everyone had. I was frequently meeting men from the British Men of War and they invited me on board of one. Unfortunately I have forgotten its name. I went as instructed, and they made a whip round and presented me with $35.00. I don't forget the Navy.

One day while walking through the Main Street I saw a man, a Peruvian, strutting down the street with my club, but not knowing the language I did not speak to him. I had forgotten to say that we brought that away with us. There is little more to record, regarding the wreck.

I might here state that Callao was not strange to me, having been there some eight years before in the old *Emigrant of Aberdeen*, and found little alteration in the place. I need say but little about my stay in Callao, as it is not a place I like anyway. I went to see the mate occasionally and found him improving. He stated that he would soon be leaving by the mail boat for England.

After about three weeks in Callao I was prepared to get away. There was a small vessel going to England who wanted three hands, so decided to take a trip in her. Her name was the *Mathewan*, belonging to Cardiff, Captain Owen, Master. Having been round Cape Horn I cannot say that I relished the idea, but such is the difference in the treatment of officers and men that poor Jack has to take his chance. This also applies to the distribution of any money that may be collected for shipwrecked men.

*   *   *   *

I have tried to find records of this interview with the British consul, but no evidence of it appears to exist in the Foreign Affairs records. After this great length of time, certain archives have been destroyed, and in 1865 there were other problems of more significance to the historian, so it is more than likely that the interview with a shipwrecked sailor, picked up from so far away, was not considered worthy of preservation.

It is unfortunate that the Consular official who interviewed Holding was not aware, or chose to ignore, The Merchant Shipping Act of 1854. "Poor Jack" did not have to take his chances, for in fact: "Consular Officers abroad are bound to send home any distressed or shipwrecked seaman, the expenses being chargeable upon the Mercantile Marine Fund."[2] So just as Dalgarno and Smith were granted immediate free passage back to England, Holding too, should have been granted the same right. It is interesting that Holding did not know of this law, for he mentions in his memoirs some of laws that did affect sailors:

At that time vessels in the guano trade used to be loaded down to the uttermost limit. In the year 1857, when I went round [The Horn] in the old Emigrant of Aberdeen, we had loaded her with 1450 tons though she was registered at 934 tons. With only three inches of freeboard we had to throw overboard 250 tons. This was the reason for the passing of what was called the Plimsol Act – the first interference on the part of the British Government, in behalf of the Seafaring Men.

We had it very rough after rounding the Horn and one day, while running before a gale, we pooped a sea and shortly after, took one on the broadsides through the man at the wheel letting the ship broach to. This just about cleaned out the galley and filled our house on deck. There was a Lady passenger on board, and poor old girl had, according to tradition, to take the blame for that. She, however, asked the Captain to sell her two bottles of rum so that she could treat us. This cleared her of all further responsibility.

After the events mentioned above, we had a reasonably fair voyage and were bound for Cowes, Isle of Wright [sic] for orders. On our arrival at Cowes, I received a very nice letter from Grearson Cole & Co. No. 1 Princess St. London, congratulating me on my safe arrival and giving me the address of my parents who had moved and also notifying me of subscriptions having been got up for us in Aberdeen as well as in London. This is now before me. I asked the Captain to allow me to leave but he would not consent, so as he was ordered to Rotterdam, I had reluctantly to go.

There is little to interest anyone in this but I might be allowed to say that when we had tied up at the Hook of Holland, it came on to blow hard. As is usual in going into harbour, we had the Union Jack on the fore royal peak; this was wound round the mast and could not be hauled down. So the Second mate asked me to go up and fetch it down. It was a most dangerous job, owing to the roll of the vessel and the weakness of the mast, for it was like the proverbial whipstock. I did however get some of it, but it was now useless.

I have now before me my discharge, which says:

conduct v.g.

ability v.g.

Robt. Holding.: Place of birth. England. Date 1840

Capacity AB. Date of Entry. July 10/65.

Date of Discharge Octbr 21/65. A. Owen, Master.

I was paid off on the 23rd and got away as soon as possible. I called on Grearson Cole and Co. who informed me that they held my share of a subscription got up on our behalf. My share of the subscriptions altogether amount to about £96.

They, like all others, treated me with the greatest of kindness.

THE END

---

[1] Her Majesty's Service – the navy
[2] *Encyclopaedia Britannica*, 9th ed. Vol. 21: p. 606

22

survivors

"Ships are but boards; sailors but men."

Shakespeare

THE first official word Mrs Barnes (wife of the Steward) had about the fate of her husband was not until 15 May 1865:

Madam,
We have now received a list of the crew that left Melbourne in the unfortunate ship Invercauld, and we find therein an Alex'r Burns [sic].

> We are,
> Madam,
> Yours respectfully,
> Richard Connon & Co.

In those days before overseas telegraph it would be considerable time before the owners would hear that their ship was overdue. Considering that the leaky *Julian* took 44 days (May 14-June 27) to travel from the Aucklands to Callao, the *Invercauld* would have been expected to make landfall in mid-June, 1864. If port officials allowed two weeks before posting her missing, the news would not have reached England for at least another six to eight weeks. Their request for the crew list, to include those nationals who signed on in Melbourne, would not have arrived there until November and it would be mid-February 1865 before a reply would have been received in England.

The first news of the rescue of the *Invercauld* survivors appeared in the *Panama Star* on 6 July 1865:

Among the passengers from Callao, by the *Chile* is Capt. George Dalgarno, late of the British ship *Invercauld* which sailed from London on the 10th of January and bound to Melbourne,

Australia. After a fine passage of 84 days to Melbourne, where she remained a month, she sailed thence for Callao in ballast on the 2nd day of May. On the night of the 10th of the same month, during a severe gale from the north ward with thick weather, she struck on the N.W. end of the Island of Auckland, and went to pieces in less than half an hour. Out of a crew of twenty-two, nineteen including the captain and officers, succeeded in reaching shore, or rather were providentially washed on to, the surf being so heavy that the exertions of the most powerful swimmer would have proved unequal to the task. Not one of the survivors succeeded in saving anything but what he had on, and not one had a pair of boots.

Returning to the wreck the following morning, all that could be obtained was some 2 lbs. of biscuit and about an equal quantity of salt pork. Fortunately there was a plentiful supply of good water on the island, to which Capt. Dalgarno assures us he owes his life, as also the lives of his brothers in misfortune, as nothing in the shape of subsistence but roots and a species of limpet (shell fish), could be found on the island. The captain could not eat the roots for the first month, and as the "limpets" were not found until he had been on the island for upwards of six weeks, the water, which certainly must have possessed peculiar nutritive properties, was the only thing that supported him.

After remaining on the island twelve months and ten days, during which time all but the captain, chief mate, Andrew Smith and one seaman, Robert Holding, died from starvation and cold, they were rescued from their perilous situation by the Peruvian ship *Julian* bound from Macao to Callao, with Chinese emigrants, which ship having sprung a leak sent a boat on shore to ascertain if they could obtain assistance. There being no inhabitants on the island but the three above-named unfortunates who were taken on board, the *Julian* proceeded on her voyage, keeping the pumps going all the time, and arrived on the 27th June, the day before the *Chile* sailed.

Captain Dalgarno and his two companions speak in the highest terms of the treatment they received from Captain C. Arrabarini and the officers of the *Julian*, and beg, publicly, to return their most grateful thanks. Mr. Andrew Smith, chief mate, and Robert Holding, seaman, remain at Callao and Captain Dalgarno is now on his way to England to report his misfortunes and providential rescue to his owners, Messrs. Richard Connon & Co. of Aberdeen, and to gladden the hearts of those relatives and friends who must have long since supposed him dead, as nothing, after the day the *Invercauld* sailed from Melbourne, had ever been heard of her until the arrival of Captain Dalgarno in Callao; just 14 months. Few more providential escapes from a fearful death have ever come under our notice, and although it is the first accident that has ever occurred to Capt. Dalgarno during his twenty seven years' experience at sea, we heartily trust that it will also be the last, and we sincerely wish him God-Speed.

On 29 July 1865, the joyous news greeted the readers of the *Aberdeen Journal* as they picked up the story.

THE LOST SHIP "INVERCAULD":

Our readers will recollect that last year the fine new ship *Invercauld*, built by Mr. Smith, Aberdeen, was lost on her voyage between some of the Australian ports and Callao. No word having been received of any of the crew, it was concluded that she had foundered at sea, and it was not expected that anything more would be heard of her or her ill-fated crew. We are happy to state, however, that yesterday Richard Connon & Co., the managing owners of the vessel, received a telegram from Captain Dalgarno, the master of the *Invercauld*, who has arrived in Southampton, stating that the vessel was totally wrecked on Auckland Island in May, 1864. There were four [sic] survivors including the captain, chief mate, and two seamen; and particulars were promised at a future occasion. The captain will probably arrive here on Monday. Of course, it is as yet a complete mystery how the four survivors did not find it possible to communicate their where-abouts all this time; and, so far as at present appears, their story will be a repetition of that of Robinson Crusoe. From the Telegraph Company we have received a telegram to the following effect: – "The Master of the British ship *Invercauld*, the only survivor of a crew of twenty-two, has been brought to Southampton by the ship *Shannon*. The ship has not been heard of for fourteen months. She was wrecked on a desert island near the Falkland Islands. Six of the crew were drowned and sixteen died of starvation on the island." The "Falkland Islands" here must surely be a mistake for Auckland Island; for it is hardly possible that the crew could have been fourteen months on the Falkland Islands without finding means of communicating with Europeans. Auckland Island is situated in the Pacific Ocean, about directly south of New Zealand, and is one of those coral islands that sailors seek to avoid; and this may account for their long and mysterious imprisonment, unknown and dead to their friends.

Four days later, the *Aberdeen Journal* (2 August 1865) produced a much more extensive report:

THE LOSS OF THE SHIP "INVERCAULD"
RESCUE OF CAPTAIN DALGARNO AND TWO OF THE CREW

It is now somewhat over eighteen months ago that we recorded the lateness, from the shipbuilding yard of Mr. John Smith, of the fine ship *Invercauld*, belonging to Messrs. Richard Connon & Co., of this city. The *Invercauld* sailed from London for Melbourne on 10th January, 1864, and in no long time after her arrival at Melbourne the *Invercauld* was reported left on her passage from that port to Callao. Since then no tidings whatever have been received of the ill-fated vessel. For a time news of the crew, or part of them, was anxiously looked for, however, and by none more anxiously than the respected owners of the *Invercauld* but for months past hope had been dead, or all but dead, in the breasts of even the most sanguine. On Friday, however, a telegram was received by Messrs. Connon & Co. from Captain Dalgarno announcing his own arrival at Southampton by the West India steamer, and the further fact that the mate and one seaman had also been saved, out of a crew of twenty-five. This telegram was precursor to another wondrous tale of the sea, of which we shall give an outline, so far as materials have come to hand. We give first:

THE CAPTAIN'S NARRATIVE,

Captain Dalgarno writes to the owners as follows:

Southampton, July, 1865

DEAR SIRS, – I am at last offered another opportunity of addressing you again in this life, to let you know the sad and melancholy tidings of the ship *Invercauld*, which became a total wreck during the night of May 10th, 1864, on the island of Auckland, off New Zealand, during a heavy gale from the northward, and thick weather. I have seen and suffered more since the disaster happened than I can pen to you at this time; but if God spares me to reach you, I will then give you all particulars. In about twenty minutes after striking she was in atoms, so heavy was the sea running, and all rocks where the disaster happened. The boys Middleton and Wilson, and four seamen were drowned, the remainder, nineteen of us, getting washed on shore through the wreck, all more or less hurt, the night being intensely dark and cold. We saved nothing but what we had on our persons, and, before being washed from the wreck, I hove off my sea boots, so as to enable me, if possible, to reach the shore.

After getting ashore amongst the rocks, we called upon each other, and all crept as close together as we could to keep ourselves warm. The spray from the sea reaching us made it one of the most dismal nights ever anyone suffered, and we were all glad when day broke on the following morning, when all who were able went towards the wreck to see what could be saved. All we found was about two pounds of biscuit, and three pounds of pork – the only food we had to divide amongst nineteen; and after all taking about a mouthful each, we went and collected a few of the most suitable pieces from the wreck to make a sort of hut to cover us from the weather, where we made a fire, the steward having saved a box of matches.

We remained four days at the wreck, and having no more food, nor appearance of getting any more at the wreck, we proceeded to go on the top of the island, to see if we could find food or any inhabitants. It was no easy matter to reach the top, it being about 2000 feet high, and almost perpendicular. When we got there we found no inhabitants and the only food we found was wild roots that grew on the island, of which we ate, and fresh water. At night we made a covering of bushes and lighting a fire crept as close together as possible. On the following morning we made towards a bay that was on the east side which occupied some days, the scrub being so heavy to walk amongst. The cook and three seamen died during this time, and all of us were getting very weak for want of food, and cold. At length we reached the bay where we found some limpets on the rocks, of which we ate heartily. We also caught two seals, and found them good food; and had we got plenty of them, no doubt all would have lived.

After living three months upon limpets they got done, and all we had again was the roots and water, seeing no more seals. By the end of August the only survivors were myself, the mate and Robert Holding, the carpenter, the boys Liddle and Lansfield, being among the last that died, all very much reduced. After we three had lingered for twelve months and ten days were at last rescued by the Portuguese ship *Julian* from Macao, towards Callao, with Chinese passengers. She sprung a leak off here, and sent a boat on shore to see if they could get their ship repaired, when they found us the only inhabitants on the island. They proceeded on their passage to Callao taking us three along with them. We were treated very kindly and on the 28th June reached Callao, where we were all treated kindly by the people there. On the same evening I sailed by mail steamer for England, leaving the mate and the seaman in Callao. On the 6th July I sailed from Panama; on the 13th arrived at St. Thomas, and sailed same day for Southampton by the steamship *Shannon*, meeting with the greatest kindness from all on board the several ships I sailed in.

I telegraphed to you immediately on our arrival at Southampton. I will call on Messrs. Grearson, Cole, & Co. in London before leaving for Aberdeen, which I hope to reach soon, and give all particulars. – I am, Dear Sir, your obedient Servant,

GEO. DALGARNO

In another letter, Captain Dalgarno writes:

Everything in and belonging to the vessel was lost, including a medal presented to me in the year 1862 by the United States Government for saving the lives of the crew of a waterlogged vessel, and a telescope presented to me by the British Government in the same year for saving the lives of the crew of a British vessel, helpless, and over which the sea was breaking in a manner which threatened her speedy destruction, both of which I valued most highly as mementos of service I had been enabled to perform to brother seamen. The island was barren of anything in the shape of food, beyond a few weeds and shell fish, which I could not at first eat and from which little nourishment could be extracted. Providentially, we found a good supply of pure water. For a considerable time after we were thrown upon the island we had not and could not provide any shelter from the piercing cold and heavy rains, and eventually we had only a poor covering constructed of seal skins which we had succeeded in obtaining from seals with difficulty and at intervals, captured by us.

They then printed a letter from the mate to his wife:

THE LOST CREW
As indicated, twenty-two lives were lost by the wreck of the *Invercauld*, six at the time of striking, and sixteen subsequently from starvation. The saved were – Captain Dalgarno, Mr.

Andrew Smith, the mate, belonging to Aberdeen, and Thomas Holding [sic], seaman. Of those who were lost and died, five only belonged to Aberdeen, whose names are – Alex. Henderson, the carpenter, a married man of very excellent character, and who has left a wife and one child; and four boys, vis: – William Middleton, George Liddle, James Lansfield, and John Wilson. The deaths of these are mentioned by Captain Dalgarno. The rest of the crew had been shipped at Melbourne, and none of them, it is believed, belonged to Aberdeen.

TREATMENT OF THE RESCUED SEAMEN AT CALLAO

It is gratifying to record that the treatment of the rescued seamen at Callao was of the kindest possible description and foremost to interest himself in their care was our townsman, Mr. J.N. Aiken. Though Captain Dalgarno was but a few hours in Callao, sum of about £100 had been subscribed to provide for the wants and pay for the passage home of himself and his companions in misfortune. Mr. Aiken, in a private note, says, "The log Captain Dalgarno has written out, of the twelve months and ten days' sufferings, is one of the most touching records I have ever read." When the long deferred means of deliverance arrived, Captain Dalgarno and the other two survivors were in a state of the greatest exhaustion – scarcely able to stand when taken on board, while, after their long and heartsickening captivity, they could scarcely believe that help had come at last. Captain D. [sic] is now much recovered, though his long-sufferings, as we understand, make him look much older.

We extract the following remarks from the Shipping Gazette of Saturday: – In the 10th of January, last year, there sailed from the port of London a ship of which nothing has since been heard until the arrival yesterday of the West India Mail. The ship in question was the *Invercauld*, belonging to Messrs. Connon & Co. of Aberdeen. She left London for Melbourne, and the fate of the vessel and the sufferings of the crew form one of those episodes in Maritime adventure which appear stranger than fiction, and the recital of which possesses all the interest peculiar to narratives of this description. Shortly before the mail left Aspinwall the *Chile* arrived at Panama from Callao, and amongst her passengers was Capt. Dalgarno, late of the *Invercauld*.

To attempt a description of the misery which Captain Dalgarno and his companions must have endured, as week after week, and month after month passed away, and no sign of deliverance presented itself, would be difficult, and, perhaps impossible. We refer to Captain Dalgarno's narrative as affording proof that the friends of missing seamen should not be precipitate in giving them up for lost, because a lengthened period may elapse without tidings being received respecting them.

The survivors from the wreck of the *Invercauld* might, perhaps, have contrived to support life for some time longer than they do on the Island of Auckland, and might, in the end have been rescued. Men who had the fortitude and the physical power of endurance to exist for twelve months and ten days on an uninhabited island, with little to subsist upon but a few roots and water, would probably have held out longer, if they had been called upon to do so.

Apart from the gratification which the restoration to their friends of the survivors of the wreck of the *Invercauld* must produce, it is always a satisfaction to have a mystery of this nature finally cleared up. There is no condition of mind so painful as suspense, and there is no suspense more distressing than that which is created by the absence of all intelligence relative to a missing ship. The adventures of Capt. Dalgarno and his companions prove, that to resign all hope after the expiration of even many months is not wise, so long as there are on the surface of the ocean uninhabited spots, where ships may perish, but where human life may providentially be preserved until the hour of rescue arrives.

A subscription has been set on foot on behalf of Captain Dalgarno, Mr. Smith, chief officer; and George [sic] Holding. The list is headed by the following subscriptions: – Messrs. Richard Connon & Co., £52 10s; Messrs. George Thompson & Co., £52 10s; Mr. William Gladstone, £30; Messrs. Grearson & Co., £21; and Messrs. J. Thompson & Co., £21.

* * * *

On August 9th, the obituaries of William Middleton and John Wilson appeared in the *Journal*. A few days later, on Wednesday 16 August, at the bottom of a column which included the notices that Mr. James F. Kean, Union Row Academy had obtained a "high place in the First Division at the recent Matriculation Examination, London University" and that the first meeting of the road Trustees was to be held on the 29th inst., we read:

The *Invercauld*: Mr. Andrew Smith, the mate of the *Invercauld*, who was left at Callao when Captain Dalgarno sailed for home, arrived at Southampton in the end of last week in good health! He writes on Saturday to say that he expects to be with his wife and family here in a few days. Mr. Smith's letter, we may add, is full of expressions of gratitude for the kindness he has experienced on all hands.

On September 6th, the *Journal* reports:

The *Invercauld*: We have seen Mr. Smith, late mate of the *Invercauld*. He is in good health, with the exception of a feeling of pain and numbness in the legs and feet. He speaks in the most grateful terms of kindness bestowed upon him by Dr. Rowe and Mr. James Aiken, Calliao; [sic], Captain Merechant, London; his owners, Messrs. Richard Connon & Co.; and indeed all he has come in contact with.

* * * *

Captain Dalgarno, to his credit, wrote a personal letter to Mrs Barnes on 25 September 1865:

Madam,

On my return here from London the other day I found Mr. Connon, owner of the late ship "Invercauld" of which I was master, had received a letter from you and that he had answered the same. I have now the painful duty to inform you that your husband died the 7th July, 1864 – and was buried beside his shipmates that shared the same fate of which you have seen the account in the papers before now. I would of wrote you sooner but did not know your address. Your husband was reduced to a great state of weakness before he died, likewise see [sic] the others that died, he was not strong when he shipped with me in Melbourne but no doubt had not the sad disaster happened he would of reached home all right. I am recovering now myself very fast but have been greatly reduced. I am sending you his pocket book that he happened to have on him at the time of the wreck, which I may say was the only article that was saved from the wreck, hoping to hear from again,

> I am
> Dear Madam,
> George Dalgarno

*   *   *   *

The final reference to the *Invercauld* is to be found in the *Aberdeen Argus*, 13 October 1865:

...His [Captain Dalgarno's] health is still very delicate, owing to the extraordinary privations to which he had been exposed, and his medical adviser has forbidden him to speak to anyone on the subject of the wreck, as any recurrence on his part to the sad event always brings on a nervous attack. Captain Dalgarno, however, states that so soon as he is sufficiently recovered it is his intention to prepare a full statement of the wreck and of the events which followed it, and this statement will be given to the public.[1]

Little else was heard of the *Invercauld*. In the opening lines of his booklet, published in 1866, the mate states: "I understood Captain Dalgarno was to give an account of them [a narrative of the wreck and sufferings of the officers and crew], and in this expectation I deferred drawing up this narrative. But as no such account has appeared, or seems likely to do so now to my knowledge, and as some of my friends and the public are anxious to possess a record of the sufferings experienced, I have been induced to bring out the following brief narrative."[2]

Perhaps this is not surprising, for George Dalgarno had sailed a new ship onto a lee shore, and like the authors of other historical blunders, a detailed account is rarely forthcoming.

Inquiries into wrecks did not begin until the 1870s and no record of any interview with Dalgarno, or report on the wreck exists in the files of the shipping company or her owners. It is as if the *Invercauld* never was – except to the families of those who served on her.

[1]   Captain Dalgarno never wrote an account. In 1866 he became master of the *Alexandria* plying the Australian and South Pacific routes. His last command was in 1885 – the *Highland Forest* to Brazil and the River Plate.
[2]   See Appendix Three for his story.
[3]   Smith, A. *The Castaways*, p.1.

EPILOGUE

*"Tourism is a valid human activity, but not necessarily compatible
with the protection of these islands."*

Department of Conservation, *Te Papa Atawhai*

THE now familiar Auckland Island wind wrapped us in a sodden blanket as once again
we searched the western slopes of Rose Island for the *Invercauld* hut site. Fourteen months
had passed since that brilliantly sunny day in December 1993 when I had commented to my
brother that I had the strangest feeling that I was not saying goodbye to the islands. I had
indeed had a glimpse of the future, for I returned in February 1995 as part of a team put
together by New Zealander Dave Asher of Southcoast Video.

We had spent two weeks filming the islands; the coastwatcher stations at Ranui Cove and
Tagua Bay, the *Grafton* and *Erlangen*[1] sites, the Victoria Passage, and the western cliffs. Our
divers had made the first dive in Lake Hinemoa; they had gone down over the graves of the
Dundonald, *Anjou*, *Grafton* and searched for the *Compadre* and *Invercauld*. Finger posts and
castaway depots, seal pups and albatrosses, shags and teals, megaherbs and rock formations
had been captured on film. Tales had been told of wartime waiting, of castaways, and of the
heartbreak of the Hardwicke settlement.

We had had our share of excitement: Mark Hammond had taken the stubby, shallow-
drafted *Geomarine* through the passage between Rose and Enderby Islands and shot through
the Victoria Passage; we had 'lost' the helicopter for a few hours and had played with the sealions on the beach in North Harbour. There had been magic private moments – standing
alone in the bush at Hardwicke sensing the struggle of the brave souls who had risked
everything for a dream – the sight of two tiny orchids on the graves of James Mahoney and
Janet Stove – being surrounded by a flock of bellbirds – an albatross rising from her nest to
proudly display her egg – seal pups playing in a nursery pool.

There had been moments of high personal challenge – landing in the dinghy in 'Invercauld
Cove' where Holding's description of the landscape seen from the shore confirmed beyond a
doubt that we had found the site of the wreck. Assorted pieces of wreckage were scattered

on the narrow, rocky beach. Two long pieces of wood with axe marks and square-headed spikes lay well up the beach. From the *Invercauld*? Unlikely, but those bits of waterlogged wood created images in the mind's eye of that night so long ago when the wreckage from a proud ship was piled high in this remote cove.

At Hardwicke we discovered a pile of bricks that were quite likely those of the chimney which Andrew Smith had spotted as the castaways had made their way down from the hills. At this site it was possible to trace the outline of the building and to find the doorstep which was hidden under a layer of peat. Despite the fact that we had photographs of the fireplace site, it was proving a difficult task to find it. From the last trip we knew where it was not – and despite Holding's estimate of the distance from the shore we were now working an area well up the ridge. I cut further westward into an area of shoulder-high tussock and suddenly something struck my leg two sharp blows followed by a long scratch. Thankfully I was wearing high leather boots and gaiters. I didn't pause to find out what it was, but I had likely been near to stepping on a penguin. DoC's Pete McClelland, who was with us again, commented: "Just be glad it wasn't a sealion." Just as we were ready to give up hope and accept the fact that the fireplace would remain an unfound relic, Karl, our sound man, shouted: "Found it!"

How could we have missed it? Six feet long and two feet high, a carefully assembled pile of flat black rocks lay nestled in the tussock. The four foot walls were gone, but it was possible to trace the indentation of the floor. Although it is forbidden to disturb historic sites, I could not resist lifting one of the rocks – my great-grandfather had probably been the last person to pick it up. Kneeling in the sodden grass as the camera rolled, I was overwhelmed with a sense of the significance of family. I thought of those who, like Mrs. Barnes, had received a letter "regretting to inform you that..." The men who had died on these islands had been loved ones: brothers, fathers, husbands, sons. They lie, scattered over the islands, unmarked and unknown. This fireplace is as close as the men of the *Invercauld* have to a memorial.

I have been incredibly fortunate to have had the opportunity to visit the Aucklands not once, but twice. They are islands where nature is wild, raw and beautiful. There is a strong sense of the tragic human history of the place for the first-time visitor. The Auckland Islands are precious because they are relatively untouched by the hand of man, and because they are a laboratory for techniques of preservation and restoration.

The 'right' of the public to visit the islands versus the need for conservation creates a moral dilemma for those charged with the management of the sub-Antarctic islands.

The term 'eco-tourism' has been coined to distinguish the business of serving up experiences centering on the ecological awareness of an area, from that of 'traditional' tourism which focuses on viewing landscape for its scenic or historic values. The argument that planned tours to fragile areas actually protect them from the irresponsible and/or ignorant tourist, does have merit, but the fact that organised tours bring more people into such an area, is an important factor when considering the long-term environmental impact of this type of visit.

On this ever-shrinking planet are those who search, for whatever motive, for ever more distant and exotic places to visit. New Zealand's pristine sub-Antarctic islands certainly fall into this category. These are not destinations for the timid: the weather can be appalling, the islands are rugged and dangerous, and getting there across some of the wildest seas in the world is difficult and expensive. However, for some, these facts make the islands even more attractive, even more of a challenge, even more of a 'trophy' in the home video library.

Wilderness tourism is self-defeating, for by the simple fact of the increased number of visitors, the wilderness which they have visited to experience is bound to be affected. Environmental protection does not stop at low impact passage, for the more difficult question of wildlife disturbance must be considered. Despite the motivation of the well-intentioned, the sub-Antarctic islands are too great an environmental treasury of rare plant and animal species to be opened for eco-tourism on anything but the most controlled basis.

Since the first tour ship arrived in the Auckland Islands in 1968, as part of a cruise to Antarctica, a quota system for visitors has been developed. In the 20 years between 1968 and 1988 only 1300 tourists visited all of the sub-Antarctic nature reserves. This number jumped to 1500 in the following six years and by November 1989, 16 applications for permits for a total of 1500 people had been received for the 1990 and 1991 seasons.[2]

Public entry is now restricted by the Natural Reserves Act in 1977 and by the guidelines set by the Management Plans of the Department of Conservation in 1987 and 1994.

The Management Objectives are stated as follows:

1. To preserve and maintain the indigenous flora and fauna, ecological associations and natural environment of the Auckland Islands in a natural state and to allow the operation of natural processes and accept their effects as far as possible.
2. To manage the reserve as an integral part of the national and international system of protected areas with emphasis on the protection of its special or unique natural features.
3. To protect and manage any biological, scenic, historic, archaeological, geological, or other scientific features to the extent compatible with the primary objective (1, as above).
4. To prevent human interference which causes undue modification or acceleration of natural processes or the alteration or destruction of natural features on the islands.
5. To allow and encourage research and studies, which will have no permanent detrimental effects especially where it has been demonstrated the results will contribute directly to the effectiveness of protection and management of the reserve.
6. To promote public appreciation and enjoyment of the features of the reserve consistent with the primary objective (1, as above).
7. To promote understanding of the specialised and vulnerable sub-Antarctic ecosystems and pursue the protection of the habitat and food source of mammals and sea birds in the surrounding sea.[3]

It was in part because of the experience of my great-grandfather and his shipmates upon these islands, that animals were introduced. It is simple enough today, in a world of instant communication and freeze-dried foods, to criticize the legislators who ordered pigs, sheep, goats, and rabbits, to be left to run wild on the islands. What was done, was done for the purpose of saving lives at a time in history when social action on behalf of the less fortunate was in its infancy. The decision would have been held as forward-thinking and an act of great benevolence. We cannot allow today's 'politically correct' movement to sweep well-intentioned efforts of the past into a manure heap of history.

However, we now have the knowledge and awareness to assess the damage which has resulted from these actions and to take steps to preserve what is left. This is the challenge facing New Zealand's Department of Conservation. Their mission is to balance the sharing of the abundance of the beauty and knowledge of these islands with the necessity, not only to keep one of the last unspoiled areas of the world in its almost pristine condition, but to make a concentrated effort to repair the damage done by the introduced species from the shipwreck and farming era and to protect this area for posterity.

It is not just the impact of humans upon these islands, with their related waste-management problems, but the attendant problems of introduced species (rodents from boats and seeds of non-native plants) which must be addressed. "Management of these island reserves is firstly about minimising the risks of visitation, which may to an extent be equated with minimising visitation... In assessing applications for visits there must be a balance struck between their degrees of risk and their degrees of benefit."[4]

Interest in these islands is certainly not the sole prerogative of the scientist, the archaeologist or the historian. Thus, an important mandate of the Department of Conservation is to bring the fragile beauty of these islands to the public view. This has been addressed by their 'off site' opportunities, most specifically the Southland Museum's display and the magnificent slide show which accompanies it. In 1990 an 'Arts in the sub-Antarctic' was sponsored and encouragement is given to film companies who can bring the islands into the livingroom through television documentaries such as the Southcoast Video project.

New Zealand has a rare treasure in these islands. Their history is a tribute to the determination of man; their present is a challenge to the Government of New Zealand and their future is in the hands of all who love the wilderness.

Madelene Ferguson Allen
Lennoxville, Que. 1997

[1] The *Erlangen* was a German ship which on the outbreak of WW II took on wood at Carnley Harbour to augment its fuel supply to enable it to reach South America.
[2] Department of Conservation, *Conservation Management Strategy*, Ver. 3: p.43.
[3] Department of Lands and Survey, *Management Plan for the Auckland Islands Nature Reserve, 1987*: p.6.
[4] ibid. p.10.

Appendix one

Early Years of Robert Holding in England

In order that the readers of this, my humble epistle, may realize and appreciate that which is to follow it would appear necessary, that I now make some few explanations regarding my reasons for undertaking the task before me. In doing so I must in consequence request the indulgence of those who favour me with their patronage.

It was decided that before publication of my account of the Wreck of the Ship Invercauld (typed to the best of my ability at the strong request of most of my family), one of my daughters and myself would pay a visit to the places of my childhood. I regret that owing to the speed at which we were obliged to travel last year [1926] it was impossible for me to see and examine a quarter of that which I wished to see.

My account will involve going into some matters at length. This will involve the mention of many names and places which will be familiar to most of those who this is expected to reach, and will at least shew that I, at 85 years, still retain most of my faculties. All of which may be considered of more or less interest to a few, a very few I fear, who have become anything near my age. It may be of some interest to many of the present generation by the recalling of many names and occurrences often spoken of to them by grandfathers and grandmothers.

At this juncture it may be excusable on my part to give the reader an idea of the state of affairs at that time as remembered by me. Going into Huntington, where the G.N. Railway now crosses, there was an old Mansion with dilapidated walls called "The Yews", (or some similar name). This was, of course, before the railway was constructed, when all the signs of the old place were obliterated and before the Cambridge line was thought of. I well remember being on the Godmanchester side and seeing the old one horse coach coming in.

My object is chiefly to show the present generation the many differences and advantages in present day conditions as compared to those existing during the times of my youth. With this in view, I will leave Huntington for the present and pass on to Brampton; and on to Ellington.

My father was a gamekeeper under Tom Knolton, who committed suicide I understand, and we then moved to Ellington Thorpe. His duties included the protection of the districts around Brampton Wood, Graffam West Wood and Coffee Wood. The latter bordered on the domain of the Duke of Manchester. My age at that time can be imagined only by the fact of non-remembrance, but I do remember the first house we lived in. This was the next one in the row now standing and next the brook or ditch. These houses adjoined that occupied by one John Hall, a "wheelwright" who sold beer to be drank [sic] only off the premises. We later moved to the centre of three houses in a field, still adjoining Hall's property. These houses are now total wrecks.

Let me name a few who lived in close proximity: First, on the hill going towards Ellington, there was and is, a row of houses standing on end onto the road with a passageway running between them.

The first on the right was occupied by a family named Sharman; the next I do not remember, then came a family named Hart, (whom I shall have occasion to allude to later). On the left was a shoemaker named Curtis and lastly a family named Wise.

Higher was living a butcher, Banks Reeves. Well do I remember that he had a widow woman named Jefferes as housekeeper who had three children. Upon one Ellington feast day, a Sunday, one of the boys reached out of the front bedroom window and set fire to the thatch roof. The ground is still vacant. He then moved to Ellington opposite the Moat.

This Moat was on the left hand side as you turned into the village from Brampton. I will now refer to a few places of memory in the Village. Going up the hill there was a grocer's shop on the left, with one or two houses; one was occupied by a returned soldier, a petitioner named Tom Wise – who once assisted my father. I have little recollection of any other buildings between that and the church yard. At the left hand side of the entrance to the churchyard, situated close to the wall, were The Stocks. To the best of my recollection there was a large oak post with means to fasten those breaking the laws and stretched along the ground were two large oak bars hewn out for the reception of the neck, arms and legs of criminals.

* * * *

In June 1993 my husband and I journeyed to the area so vividly described by Holding, to begin our verification of the manuscript. We wound our way through lovely rolling farm country of Cambridgeshire and came into Ellington through Ellington Thorpe, which brought us in at the bottom of the long hill as described. We started our tour at the church, near to where we thought the stocks had been located, and wandered in the graveyard where we found a few tombstones with names from Holding's account. Leaving the car in the little square in front of the post office/grocery/bus shelter we proceeded down the hill to try to find the moat. After tromping up and down and seeing nothing, I asked a woman who was letting a couple of little children out of a car in front of an old thatched cottage if she knew of an old moat nearby. She replied that she was only chauffeuring the children but suggested that I speak to the "lady of the house". This woman very kindly invited me in to her historic home (400 years old, as it turned out). The Wallises provided a wealth of information, for their home and The Mermaid pub were the two oldest buildings in the village. The grocery which Holding refers to had been in this same building. They showed me the old deeds of the house, which had at one time been owned by the Duke of Manchester.

Meanwhile, Robin had disappeared into the yard of a private home in the area where the moat should have been. He found an overgrown indentation in the ground which must have been the old moat of the now long gone manor house.

There is some confusion as to Fish Pond, for according to Mr Wallis, the area opposite the green had been a fish pond but had been filled in. However, the map of Kimbolton shows a large fish pond to the east.

* * * *

On the opposite side of the entrance there lived a family of the Peculiar People or Latter Day Saints. Referring back to the other side of the street there was a shoemaker shop, then a lane running to the brook, then a large new bakeshop occupied by Jim Measures. There was little on the left with the exception of a public house, until we come to the tollgate, which is now demolished, with the opposite kept by one Knowles. The Clergyman named Ludman lived on the left going to Ellington Thorpe and little cared for his flock.

Across the bridge was the entrance to the Duke of Manchester's stables and turning to the right and shortly to the left up to what was once called Narrow Street. On the left hand corner was our grocery, kept by one Brewster, and next him on the left was Dr. Fernie, who had three sons drowned in the fish pond. (Two of these I saw taken out.) Next in line was Mr. W. Ellis, who was Mrs. Flander's father; then the Wesleyan Chapel, the bullyard and the pump, with three houses as they stand today. Facing the street was a large grocery and on the left, old Mr. Tom Flanders yard, with several houses at the back and the Moravian Chapel built, I believe, by Mr. Flanders. In the front, going back, we come to the Bakehouse owned and worked by him with his brother, poor Old Ben, as his baker and Jacob Hinch as his help. Then came the old Britain Buildings where we stayed for two nights last June. Next, the Tanyards at the back which was then occupied, (1853-54), by Charlie Flood who later married Mrs. Tom Flanders.

* * * *

After a morning spent in the county record office in Huntington we realised that a good deal of Holding's account reflects the area as he remembered it after he returned from the Aucklands; that is, in 1865, not in the early 1850s. Through census data we discovered that when he spoke of "old Ben, the baker" he was referring to an apprentice of 16 in 1851 – it was his father who was the baker. However, he was correct for most families. We can forgive him mixing a generation here and there.

* * * *

We will now turn up to the right and past a house kept by one West; then the malt house and on to the corner turning into Front Street. Then comes the George Inn, then Flanders, then Joe Clare, watchmaker, then a plumber and next a tailor, then a grocery and a beer hall, then another called The Saddle, and so on throughout. There was one just beyond the Bullshead, which may be more interesting in relating this story than can be imagined at present. We will move on to the entrance to the Church where Mr. Stergeon had his drug store and as it stands to this day. Opposite him, going towards the Narrow St., lived Tom & Harry Bloodworth. Turning back we saw the Corn exchange which has now entirely disappeared. This was once used as a market place where the farmers used to bring in their produce and beneficial to the population it was for that purpose. But having had a wide experience in different parts I have reluctantly come to the conclusion that people do not have the energy to appreciate conditions existing.

I well remember the construction of the Gas works there and that it was thought at the time, their enterprises would ensue, but nothing was seen to have been accomplished at that time. We will now go through the churchyard and return around the Grammar school. There was to the right a blacksmith shop kept by Ibbes, I believe, and then we get to the entrance to the school. The Vicarage occupied by Mr. Axworth; the veterinary surgery next Hadfield's Pub and another next door. How all these were able to pay the licence fees is a mystery to me.

Having proved to my satisfaction that my memory is still very good I needn't go further into details to people living there at the time and will now confine myself to a few incidents which are likely to remain in my mind during life. In front of the Flanders was the entrance to Kimbolton Castle with green swab on each side of the drive and a wall with iron stanchions and chains on top partly around it. One time I was playing marbles with one of the Darlow sons with the eldest looking on; he took the marbles which I had won and wouldn't give them back. I wanted them back and he wanted to go to the green and fight. When we got there he threw a small piece of rock which struck me on the body, I naturally picked it up and threw it back, striking him on the crown of the head. He turned round and fell stiff. Someone fetched the constable, John Hadfield, who carried him home and then came back to caution me as to the result if I had killed him.

There was about that time one other incident worthy of record. The George cellar was broken into and five gallons of liquors were stolen. The Willis brothers were apprehended by John Hadfield and when they stood trial Jim got seven years and Jack was acquitted. Jim went to Australia and I met him many years later.

I well remember hearing my father deplore of the untimely death of poor Lamb, "a policeman who had been thrown over the bridge crossing the river". This later was apparently cleared up, some few years since, by the confession, on his death bed, of the man who threw Lamb over.

We will now get back to Ellington Thorpe. The life of a gamekeeper at no time was a bed of roses and while living in the house alluded to, there were many occurrences which caused us much trouble and annoyance. The family was increasing at one every two years and me about 8 years. My father's wages at that time amounted to the magnificent sum of 9 shillings per week. Owing to the Crimean War living was scarcely possible with everything so dear. Bread was $11\frac{1}{2}$ cents a loaf and we were making up half wheat and half barley flour and selling it at 13 pence per 3 lb. loaf. There was in consequence much poaching both by snaring and night shooting. It can therefore be readily understood that we as children had to meet the likes and dislikes of those that way inclined and take our share of the insults intended for father. These came not only from those in the immediate neighbourhood but from the villages within a radius of several miles.

To illustrate I will give a few instances which occur to me as I write. Charley Hull and Bill Chandler of Bugden, were brought home by my father night poaching and they threatened to shoot father. These two were later given seven years for sheep shooting. One night father and his assistant, both unarmed, were driven home by seven men with blackened faces. When they got to the house and father had got his gun they dared not come any nearer than the bridge crossing the narrow brook.

Then there was one Ike Oddle from Brampton who had received three months for poaching. He came over one dark night with the sole intention of shooting father. He came into the house while we were at supper, mother having taken up the argument on behalf of father, he struck at father across the table. I will never forget how quickly he was outside and running down the pebble walk; he missed the bridge and fell headlong into the brook but he was quickly fetched out by father and given what he deserved. As well as getting locked up and he got three months at Huntington.

Much of what I learned was of later value to me in regard to snaring as will ultimately be of much interest to the reader of The Wreck of The Invercauld. From the age of seven I was called upon by my father for assistance by helping to make snares and also to help in setting them in the evening and going out in the morning to fetch our catch. Frequently I had been fetched out of bed in the dead of winter between four and five in the morning, when the ground was white with frosts, to walk to Brampton Wood to lift the rabbits and snares. I have often heard my father say his heart ached many times for my condition. On one occasion we caught thirty-two, so that it will be acknowledged that we had enough weight to carry from Brampton Wood to Ellington. This was chiefly for the benefit of Tom Knolton and the game dealers of Huntington. I learned that this was the cause of his losing his position, also the Woodman, John Hammond who took a public house at Brampton where I stayed about Christmas time in the year 1865.

Although there are many things of the kind I could mention, I will only bore the reader with one more case which occurred when I was about 9 years old. One John Bellomy, whose home was Bugden, assisted father at this time. Father had been out several nights watching for poachers and upon this occasion requested that Bellomy take me and my brother Bill, two years younger, and go across the field to a stable. He instructed us to keep awake and listen for any shooting. To our surprise at about 12 o'clock we did hear a shot in the direction of Coffee Wood. Needless to say we were quickly on the run to inform him; he then sent one in one direction and one in another for assistance. We soon mustered four men and myself, my brother being left at home. Our object was Graffam West Wood where we went up about the centre and waited.

It was a beautiful moonlight night and very quiet and still until we saw the flash of a gun and heard the report, as well as the fall of a pheasant not more than 50 yards distance. We were on the opposite side of that which we called a riding. Father picked up his gun and ran across, went on one knee and waited for their coming. He declared that the foremost had fired the gun three times. He then sprang up and without putting his gun to his shoulder discharged it in their direction and wounded both slightly. Of course there was a fight. The first, being a big fellow named Bill Low, butted father into the bush, while Bellomy knocked down the next, a Peter Hart. Seeing father getting the worst of his encounter, I told him to go and look after father while I, at about 9 years, would take care of Hart. We took them home and well do I remember they helped to clean up a fine green ham cooked only the day before. While they were sitting there I pointed out to father that Hart had the tail of a pheasant sticking out of a hole in the corner of his pocket. Low got 18 months and Hart 6 months. Low lived next door to my uncle, who was with us that night and Hart lived, as before described, on the hill. I had occasion to write him later from Australia but he never forgave me.

There appears little more to be said regarding Ellington except that on my last visit it appeared to have progressed backwards and to shew the impoverished state of the people at the times alluded to. Money was a rarity and to get a few coppers we had to resort to all sorts of dodges. Often when we killed a carrion crow which I sold for food to an old woman living next door, whom we designated as Nanny Revels. We often killed sparrows to sell at a farm on the right hand side as you turn to Spaldwick. At one time I walked to Wolley to pull weeds from among the pebbles in a farrier's yard there at 1 shilling per week. Later we walked to Spaldwick to school under Jim Nurish, to whom my father paid 13 shillings for one quarter's education and is therefore any excuse for my lack of education.

We will now refer more particularly to Kimbolton and incidents leading up to the eventful experiences of a strenuous life.

My father had taken a position previously occupied by one Barbour at "The Warren", under

Tom Bollard as head keeper to then, Duke of Manchester. His wages were, I believe, 12 shillings per week. This made things somewhat better, but the family was still on the increase and it was difficult to make both ends meet at times and there was no allowance for the assistance which I was called upon to render which consisted of trapping and snaring. Our family at the time alluded to consisted of Father, Mother, myself as the oldest, then William, George, John, Joseph, Fannie, and Eliza, but in the interval of my apprentices we had at times to live in a hard way but always had sufficient to eat.

My schooling had been very much neglected, there being only one school and the schoolmaster was a Benjamin Clarke of Stanley, who was the terror of all his pupils. When in Kimbolton last June I was shown a large ruler with which he had hit my brother George over the head. There was a boy living in the church yard whose ear he had put out of its socket. Can there be any wonder why children abhorred to go to his school, or the want of education, as in my case?

I will now try to enumerate some of the different names of my memory. In Kimbolton, first came "The Rook", which was amongst the first houses, where lived Jack Darlow, a rat catcher with a large family of children and dogs and one old donkey. Opposite their alley was, and is, the remains of a cart house with the roof now missing. I used to have to put up our cart here. The Darlow boys were often very annoying. The third had been very busy along this line, when he ran in the house and picked up a canister of gun powder and threw it into the fire with the result that his face was terribly disfigured. He never troubled me again.

At the back of the cart shed, facing the bridge was a row of perhaps four or five houses. The first was occupied by the Newmans, I believe, then came the Willises – Jim and Jack the former I may have a few words to say about later. The names of the others have slipped my memory.

A job was found for me in St. Neots, for one Johnnie Rutter which might be designated a "ROTTER" for the reason that he wouldn't attend to his business but spent most of his evenings playing dominoes in a pub leaving me to do the work. I had the setting of the sponges and at my age to fetch down from the loft and carry round two sacks of flour for that purpose. He had a son older than I who never did anything and this, together with being half starved, I thought three months sufficient for that purpose. I had often after breakfast gone and stolen a penny loaf to appease my hunger.

*   *   *   *

Kimbolton is approached by a highway which takes an 'S' bend (actually more of a 'Z') with high brick walls inset with great gates on one side. Once this is negotiated, you are looking down the main street. To our left was Kimbolton Castle; the green in front was just as described. We learned that 'The Warren' was still standing and that a child who had been born there, to a gamekeeper's family, had just died at 90.

Following careful directions, we found ourselves just outside of the town, looking at a narrow, two-storey, red-brick building standing against the skyline. We hiked along a dirt track and then turned onto a tiny woodland path (with too many nettles on either side) which came out beside a field. Clambering over the furrows, we finally reached 'The Warren'. It stood starkly behind a formidable six foot (1.8m) high wire fence topped with barbed wire. The windows were bricked over, so even though the gate at the back was open, we decided that it wasn't worth the effort to struggle through waist-high nettles as the door would surely be locked. It stood completely alone – no road, no houses – nothing except a fantastic view across the countryside and down into the village. It was a perfect site for a gamekeeper's home.

Back in town in the Church Close we visited a Mr Stratford, the local historian. I left a copy of my manuscript with him, and several weeks later received a copy of a map of Kimbolton and a page of notations. Brewster lived at 1 East Street; Dr. Fernie at number 5; the Wesleyan Chapel later became the Oddfellow's Hall. "Then a plumber" – John Wimpers, "next a tailor" – possibly Robinson, "then a grocery" – Joseph Baines, today Mayers grocery, "and a beer-hall" – the New Sun public house today. The Bullshead was for some years known as the White Lion and is now The Cromwell House. And so it went throughout the village. Robert Holding's memory had been virtually flawless.

* * * *

My brother William had been apprenticed in London but still the family was large and another job had to be found. This turned up at Weldon in the Woods with one William Carley and for reasons I need not mention, I only stayed there about three months. When I returned to Kimbolton I worked for Mr. Flanders. All found my wages were 3 shillings a week and living of the best, but for some unaccountable reason I was still somewhat unsettled, perhaps it may have been older associates. However, that may be, the fact remains that upon one Sunday afternoon, about four of us, having nothing to draw our attention got up into the park near Tom Ballard's and began to play pitch and toss. When I won some shillings, the conversation turned on going to London. It was quickly arranged that Charley Ellis, Tom Newman and myself started to walk to St. Neots to catch the train for London where we arrived in the evening. I had been up to the Exhibition of 1851 and had the address of my brother and ultimately went there to live and polish boots. The others having exhausted their funds, walked back to Kimbolton.

It is my opinion that the foregoing is sufficient to show how easily the course of one's life may be changed when young, but the events of little report. I am sure that there are bitters as well as sweets in any hasty undertaking of the kind here alluded to. This was proven to me up to the hilt on the 3rd day of December in Gravesend on my way to India. By the recording of the foregoing it is the wish and hope of the writer that there is nothing herein contained of an unpleasant nature or in any way offensive to anyone who may happen to be living or to their offspring or any connections.

I now regret that when in Kimbolton last June my time was very limited and will conclude by stating that the information of those who have knowledge of the family that William died in Australia many years since, George, who had become notorious as a crack Exhibition shot, winning many medals under the nom de plume of Captain G.H. Fowler and who was an instructor of Ammunition Chargers at Little Stukely during the war, died on the third day of May, 1921 and was buried at Godmanchester. John, who became a manufacturer at Sheffield, died there last September after my visit. Joseph died some five years since at Southport. Leaving Elizabeth and myself as the sole survivors off all connected with the foregoing, but an immense number of Holdings throughout the Universe.

I have the honour to be yours faithfully and respectfully,

Robert Holding
Chapleau, Ont. 1926

APPENDIX TWO

Robert Holding in Australia

I had had fair bush experience, having travelled from Two-Fold Bay to the Snowy River diggings, thence through Adelong and on to the Billibung Creek, through to Deneliquin on the Mowina and thence to Bendigo and Castlemaine, Back Creek and Creswicks Creek. I think I can claim a fair knowledge of Victoria. All this may appear out of place here, but it will be seen later that the foregoing was undoubtedly the saving of other lives besides my own.

At one time I worked as a bullock driver. I had been working for a hotel keeper by the name of William Nutt in Back Creek who had a contract with a stamp mill. As he was not very prompt in paying his workers I got tired of cutting and sold out, rather cheaply, to my two partners. When I went to Billy Nutt for my back wages, he asked me to drive his bullock team of four animals, so I undertook the job at one pound per week and found.[1]

It may be explained that when yoking up bullocks, it is necessary to have a bar of hardwood about three inches square, hollowed out for the neck of each animal and long enough to reach over their necks. An iron bow made of 7/8 round iron is pushed up from under the neck near the shoulder and passed through two holes in the yoke. One then reaches over the neck of the near side animal to couple the off side one by pushing two split pins through the holes in the yoke. The process is repeated with the nearside one.

Well, the nearside one was a brute, for he would either butt with his horns or kick with his feet. The food likely had a great deal to do with their temperament for they were allowed nothing but what they could pick up at night and anyone who has ever been in Australia will acknowledge that it would mean starvation in a very short time. Now comes the result.

My own food was equally deficient and of inferior quality, so much so that I decided to throw up my job. I, of course, took my meals (that is breakfast and supper) in the kitchen behind the bar. This night there was a scarcity of meat and sugar and the bread was dry. When I asked the servant for more meat and sugar she told me that the Misses wouldn't give her any more, so I picked up my plate and went into the bar where both of them were in front of the bar with some company. I inquired if they intended to feed me on dry bread. Nutt fetched the bread and stormed about my use of the words, "dry bread". I told him what I called dry bread was bread without meat or butter. The latter I had never tasted there! To cut it short, I quit right there and got no pay for what I had done for them. I left the bullocks coupled up, never intending to see them again. I can now see that it was a cruel thing to do but I intended to punish the bullock for his kicking, and the man for his treatment by starving me. The next summer I learned that the poor bullocks had gone to a hole to drink and had fallen in head first and were found dead there. So I think I got all that was coming to me.

---

[1] Meals and living expenses.

APPENDIX THREE

THE CASTAWAYS:

*A Narrative of The Wreck and sufferings of the Officers and Crew
of the Ship "Invercauld", of Aberdeen on the Auckland Islands,*

By Andrew Smith, First Mate, 1866.

INTRODUCTION

Having been requested to give a narrative of the wreck and sufferings of the officers and crew of the "Invercauld" of Aberdeen, I have to state, as some excuse for the delay in its appearance, that I understood Captain Dalgarno was to give an account of them, and in this expectation I deferred drawing up this narrative. But as no such account has appeared, or seems likely to do so now to my knowledge, and as some of my friends and the public are anxious to possess a record of the sufferings experienced, I have been induced to bring out the following brief narrative. I must apologise for its numerous and serious imperfections. Had I possessed any materials for keeping a log, this narrative might have been much extended, but not having had anything to keep a record by, I am unable to give the date of any particular event, which I regret. Captain Dalgarno had a few scanty notes, kept on some strips of the margin of a Melbourne newspaper, but these I have not even had the advantage of seeing. I therefore write entirely from memory, and may have omitted many materials of interest, and possibly mixed up some things not in their proper order.

Such as my narrative is, I here submit it, and while doing so must express my warmest gratitude to all who showed me so much genuine kindness after my rescue, both in Callao and on my passage home from that place, and, indeed, I may say to all with whom I have come in contact since coming home.

NARRATIVE

We left the port of Melbourne, on the 2nd day of May, 1864, in ballast, with a crew of twenty-five men, all told, bound for Callao, to take in a cargo there for England.

When we left the weather was very fine, but during the night it commenced to blow hard, and after a good deal of trouble and fatigue, we succeeded in getting safely through Bass's Straits; at all times a very dangerous passage, owing to the great number of sunk rocks and shoals that beset the passage. The wind still continued to blow very strong, with showers of sleet and snow, for some days, until we sighted the Auckland Islands on the 10th of May, the gale still increasing, with very heavy seas and great showers of rain. We were running under the three close-reefed topsails, when the man on the look-out reported land ahead. We were at that time close to the Islands, and thought, but were not sure, that this land was the south end of the Auckland group, so we brought the ship a little more to the south, in the expectation that we were clear of danger. We had not run long, when to our wonder and astonishment, land was again reported right ahead, so we had no other resource than to bring the ship to the northward, which we did, and made all possible sail that we could under the circumstances, thinking to get a passage between the small island that we sighted first and the larger one. There appeared, however, so many rocks, reefs and breakers ahead, that we saw it would be very dangerous, but still carried on sail, in hopes of getting through this passage, as we knew there was no other chance of getting clear, owing to the direction from which the wind was blowing. It was then very dark, with heavy rain, a hurricane blowing, and a tremendous sea running and any one who knows about beating a light ship off a lee shore can easily understand what our thoughts were when we expected every minute to strike.

What a dismal appearance it had when we looked at the immense cliffs looming above us, and on which we were momentarily expecting to be dashed to pieces; but still we carried on all the sail we could, amidst all the dangers that surrounded us, in hopes of getting through the passage safely.

At last we got amongst the breakers, where we at once lost all command of the ship. Had we struck two hundred yards on either side, not a man would have been saved; but very fortunately for us we got driven in between two very high cliffs broadside on. One can have no idea of the height of the cliffs, but we were breaking the royal masts on the rocks above us before the hull of the ship was driven on shore as far as it could get. It would be utterly impossible to describe the scene that followed after the ship struck. It was pitch dark, the surf was running very high, and the seas making a continual break over us. We all threw off our boots and heavy clothes, as we saw that we must trust to being washed ashore as the vessel was fast breaking up under our feet. After being tossed and bruised amongst the fragments of the wreck, I was thrown on a narrow ledge of rocks, where some time elapsed before I met in with any of the crew, owing to the darkness of the night and the large boulders of rocks amongst which we were. When at last a number of us met on the shore, we huddled close together, endeavouring to keep ourselves warm, for it was the dead of winter there: the cold was intense, and of course we were all as wet as could be. I cannot describe the sufferings that we experienced, through cold and wet, the first night we were on the island. We had no shelter, and the sea-spray, with rain and snow, beat upon us the whole night. Towards morning we managed to make a sort of shelter, or break-wind, for ourselves, of part of the wreck, which we had very little difficulty in getting, as it was always washing up among the rocks. The deck plans were the most serviceable, and with them we contrived to make a little shelter from the piercing cold and wind. We waited anxiously for daylight, not knowing how many of us were saved.

In the morning we found that six of our number had perished, and nineteen had got on to the rocks. I cannot describe what were our thoughts when we saw the miserable and hopeless condition we were in. There was very little of the ship to be seen; and with the cliffs towering apparently some thousands of feet above us, we hardly knew what to do.

The steward had a box of lucifer matches in his pocket. They were, of course, all wet, but after a good deal of trouble we got some of them to light, which was very fortunate, for we were in a most miserable condition from the cold. We were also beginning to feel the pangs of hunger, and were always in expectation of getting something from the wreck, but I am sorry to say that we got nothing of importance, except about two pounds of pork and about as much bread. In this miserable condition we continued for five days. We got some water off the rocks, and were very thankful for it, even although it was very brackish. We were always crawling about among the rocks in search of something to eat. We were all extremely hungry by this time, as we could only get some wild roots which grew on the upper parts of the rocks. These roots we ate heartily, not knowing whether they were good as food or not, and little caring, as long as they for a time satisfied our hunger.

About this time, some of our party having succeeded in reaching the top of the cliffs, returned and reported having seen a great many tracks of wild pigs, one of which they had managed to kill. It was a very small one, but the news was most acceptable to us. The remainder of the sufferers all gained the top of the cliffs, with the exception of one poor fellow. He left us previous to our starting from the place where we were wrecked, to look out for some place to try to ascent the precipice, and it would appear that he had fallen among the rocks and cut and bruised himself to such an extent, that when he returned to us he became insensible. We had to leave him, as he was unable to undertake the task of climbing the immense cliffs. One of our number had the kindness to volunteer to stop with him until death should put an end to his sufferings, for he was so much bruised that we were all sure it would not be long before he died. It was heart-rending to think that we could give him no help, while the same fate was staring us all in the face. We shook hands with him before we left, and bade him a mournful good-bye.

We then commenced our tedious ascent of the cliffs, having left the poor fellow who volunteered to stay with the dying man to follow; and I am thankful to say we all succeeded in reaching the top. It was frightful to look back and see the steep and perilous way we had come. Every now and then the long grass gave way with us, and at times large stones would get detached from the rocks and roll down, to the imminent danger of those below.

The time required to get to the top occupied, I think, the greater part of a day, but as I had no watch, neither had any of the others, I could not speak exactly of the actual time thus occupied. We were destitute of everything; some of us without shoes or jackets; and as the wind and rain were as cold and piercing as ever, one may imagine what a miserable condition we were in, and what our thoughts were, not knowing what was before us. But our spirits rose when we saw the party who had ascended before us and whom I have previously mentioned. They were coming in our direction with the young pig, which I have already stated they had killed. They had managed to catch only this single one out of a flock of young ones, by getting it separated from the rest.

We then commenced our tedious march through rank grass, so rank and high that we were hardly able to force our way through it in some parts. Our object was to find some place of shelter to camp in during the night. We at last found a place somewhat sheltered, and there we made a sort of break-wind. We then lighted a fire and roasted some of the young pig, which we found to be very palatable. Indeed the flesh of it was remarkably sweet. We gave it little cooking for fear of losing too much of it, and wasted nothing; some of us even lapped the blood off the ground. After supper we made up a large fire for the night to keep ourselves as warm as we could, for it was very wet and cold. We were always wishing for day to break, as the nights at this season there were long and dreary, and particularly so to us.

The next day the man who stayed with the one whom we left dying joined us. He found us out by following our tracks, and reported that the poor fellow died in the course of the day after we left them. I am sorry to say that I cannot remember their names, as most of our crew were new hands, shipped in Melbourne immediately before we left that port.

Day after day we continued our wearisome march through the long tangled grass and scrub, in hopes of reaching the east end of the island, which had the appearance of being more sheltered from the heavy gales of wind which were continually blowing on us, accompanied by tremendous showers of hail and snow.

The second morning that we were on the top of the island came in a little more moderately, for which we were very thankful. We took a little more of the pig, and there being a good supply of wild roots, we ate heartily of them, as they were a good substitute for bread. After breakfast we started to push on towards the east end of the island; a party of four separating from us to search for wild pigs. They, knowing about how far we would go, were to join us in the evening, and we were to have a fire ready for cooking, if they should be so fortunate as to catch any pigs, and also to keep ourselves warm during the night. This day's march was a very weary one, and as some were very much reduced in strength it was very hard on them. Those who were strongest would push ahead of the others for some distance then sit down and rest, and eat the wild berries, which were very plentiful, until such time as the others would come up with them. There was one kind of the berries very bitter and another very sweet, but with them and the roots we managed to keep hunger somewhat away, and found plenty of water in holes or running streams. We continued thus our weary march until nearly dark, and then we commenced our usual evening's work by collecting firewood and making a break-wind to shelter us from the piercing cold and heavy showers. Of the four men who went in search of pigs only three joined us. The man whom they left was so much reduced in strength that he was unable to keep up with them, so he sat down. It was useless for us to go in search of him that night. It was pitch dark and the scrub was so thick that it would have been impossible to have found him. So we huddled together and passed the night as best we could. In the morning we went to search for the missing man, and when we found him he was quite dead, poor fellow. He was cook, and joined the ship in London, but his name I cannot recollect. The other three who went in search of pigs said that they had seen none. They had heard plenty of them among the scrub, however, but it was so thick that they were not able to follow them.

That night we finished the pig we had caught, and there was only a very small piece to each of us. We could never get another although there were apparently a great number of them about the place. We could hear them among the scrub, but it was so dense that we could not follow their tracks amongst it. We saw some very large ones on the face of the hill on which we were camped, but could never get near them; it was so difficult walking in some parts.

From this place we had a good view of the sea, and the other islands of the group. They appeared to be a great distance off, but it was the fogs that we had at times, and our great height above the level of the sea, that made them appear so far away from us.

After a long weary march we came to the end of the mountain which looked down to the sea, of which we had an extensive view. The scrub between us and the beach lay apparently very dense.

We camped here for some days, in hopes that we would see some ship passing. We kept up a large fire to keep ourselves warm and as a signal should any ship come in sight. But day after day, and night after night passed wearily on, and no appearance of any ship coming to our rescue.

We were very uncomfortable in this place, owing to its exposed position, while there were only a few bushes to shelter us from the heavy showers of hail and rain that were continually falling, and the wind was still blowing very hard.

During our stay here six men made up their minds to go back to the wreck, which they did. Holding, who was rescued along with the captain and myself, was one of the number, but he came back to us some days after, and reported that nothing had come ashore from the wreck, and no appearance of anything coming. He considered therefore, that there was no use in staying any longer, and wanted the others to return with him, which they declined to do, though he urged upon them that if they stayed long at the wreck they would be sure to die, as there were no shellfish on the rocks, and the wild roots that we ate were very scarce.

By this time we could see no other help but to make for the beach, where we were sure we should find some shell-fish and we were in hopes that we might catch some seals. It appeared almost impossible that we should be able to make our way through the thick scrub in our then weak condition; but that we must try was evident, as the roots were getting scarce, and we now had to go a good way for them. Fortunately we had a good supply of water pretty close to where we were camped.

At last we made up our minds that we would try to get to the beach, whatever the consequences might be, as there was nothing but starvation and death to be expected if we remained where we were. It was proposed that half of us should go down first, accordingly half of our number started one morning. I am glad to say that we got down much easier than we expected. We got to the beach sometime before it was dark. How glad we were to see so many limpets on the rocks, and how thankful we were of them I need hardly say. They were more palatable than the wild roots, although we took part of both. We went about two miles among the rocks, eating limpets and roots, until we came to a place where we could camp for the night. On our way we killed a sort of bird called a widgeon, with which we made a very fine supper, by roasting it on the fire. After supper we made ourselves a little shelter, but we had a very bad night of rain. We were, however, pretty well protected from the cold wind that was blowing.

At this time we had an idea that some sealer or whaler might touch at the islands during the sealing or whaling season; some cheered themselves with the thought that we should be able to maintain a lingering existence, at least until then.

The day after we got to the beach one of our party went back to the others with some shellfish, and the joyful tidings that there were a great many of them on the rocks. They had little trouble in getting down to us, as we had broken the bushes on our way down. We expected them to join us on the day after we sent the man back to them, but it was the second day after until they came, as two of the poor fellows had died during the night after the man went back.

We were now reduced to ten in all. Up to the time that we got to the rocks – I think about twenty days after the wreck – we had had nothing to live upon except the small piece of pork and the hand-ful of bread which came from the wreck, the small pig which we had killed, and some wild roots.

We lived at this place for some time, until we had the rocks cleaned of the shells. We had during that time gained a little strength, as the limpets were very full and in good condition. After eating all the limpets we could find we proposed to go to the next bay, so a party of five started for it. We had a great quantity of bush and scrub to go through. On the bushes we found great numbers of wild berries, of which we ate heartily. We had no difficulty in getting plenty of water. On gaining the top of the hill which overlooked the bay to which we were going, we were cheered beyond expression, and leapt for joy at the sight of two houses close to the beach. With what strength we had left, we ran eagerly through the scrub to get to them, thinking that they might be inhabited. But, alas, how great was our disappointment, for when we got to them we found them quite in ruins. Still, we were very thankful for the shelter that they would afford us, even in the condition in which they were. We could not keep ourselves from the heavy rains and snow that fell at times; and we forthwith commenced to clear out one of them. As they were all growing with rank grass and weeds. Two of our party went that night along the rocks to see what they could find, and they were so fortunate as to kill a seal, the greater part of which we carried to the house that night. We made ourselves as comfortable as possible, and had a part of the seal for supper, – the flesh, we found, was very good. After supper we commenced to look about the houses where we found a good many preserved-meat tins, which became of great service to us, as we used them to boil the limpets and seal flesh in. We also found a piece of sheet-iron, of which we made a frying pan; and an old worn-out adze and a hatchet, which were invaluable to us in cutting the firewood. Before this we had to seize hold of the branches and break them off. The adze and hatchet therefore saved us a good deal of trouble and wearying work.

The party that we left soon joined us and took up their abode in the other house, which was close to the one in which we were living. We gave them a share of the seal we had caught, and as long as it lasted we found ourselves gaining strength day by day, but it did not last long, and when it was finished we had to take to our old style of living on roots and limpets. It was a good while until we got another seal.

Anxiously, day after day, did we look out in hopes of seeing some ship making her appearance, but in vain, though it was evident that a great deal of work had been carried on here at one time. We saw the places where the people had boiled the seal oil, and great cuttings through the scrub which they had used for firewood.

Having found a long spar on the beach we cut it up and made a raft of it. The raft was of great use to us. It enabled us to get along by the rocks, there being no swell in the small bays. This saved us the trouble of travelling among the rocks when we were in search of shell-fish and seals. Sometime after our first seal had been finished, Henderson, the carpenter, and Mahoney, second mate, were out with the raft and were fortunate enough to get another seal about a mile from where we were living. We got it on to the raft for conveyance and heartily wished we could get such a cargo frequently. It was a fine cow-seal. The flesh of the cow, it may be remarked, is much sweeter and more palatable than that of the bull; and besides, we never had much trouble in killing the cows or young ones, but we had some terrible fights with the bulls. The flesh of the young ones is very sweet. Unfortunately for us, seals, old and young, were very scarce at this part of the island, and after this time we could never see any more in the water, nor even any of their tracks on shore.

From the spot where we lived we had a good view of the sea to the eastward, and many an anxious glance did we cast over the desolate waters for any passing ship, but still we looked in vain. Things were assuming a very gloomy aspect indeed.

We had too, the misfortune to lose our raft. Having little or nothing to make it fast with – only a piece of sealskin – it went adrift, and we did not recover it. Before we lost the raft, four of us went with it to the east end of the island, a distance of about four miles. We went in search of seals and shell-fish, as by this time the shell-fish were getting very scarce near where we were living. While at the east end of the island we could get plenty, and it seemed to us that we would have more chance of getting some seals there. From that point we had also a better view of the sea. By this time we were all getting again very weak. We had had no seal for some time, and some of us were so feeble that we could hardly crawl among the rocks in search of limpets. Our condition, as may well be imagined, was most miserable; our clothes were all very much torn, and at that time it was bitterly cold. Some days we had very fine weather, but in general we had heavy gales from S.W. with great falls of rain and snow. This was, I think about the month of July, and it was about this time that the steward and one of the boys died.

For the reasons just stated, we thought it would be better for us to go to the east end of the island; yet we were very reluctant to leave the comparatively good shelter we had. Holding thought he would go himself, which he did, and returned sometime after. His report on his return was that he could find plenty of shell-fish with little trouble. He tried to persuade us all to go with him, but all declined to go, with the exception of myself. Previous to his return to us, however, two others had started to go to the east end. Holding and I set out, and on our way met them returning home again. They had failed in finding their way. We advised them to turn again and go with us, which they did. On our way we went down to a small bay to get a few limpets to satisfy our hunger, which was getting very keen. We had a good track the greater part of the way down, the trees having been cut down at one time for some purpose or another. When we got as far down as we wanted, we got among the rocks to look for limpets. We had not been very long amongst them when we found a large cow-seal, which we had very little trouble in killing, although we had no clubs or anything of the sort at the time. Luckily, the seal was asleep, and arming ourselves each with a stone, we crept up to her very quietly and struck her on the head, which stunned her at once. Thanking God for his great kindness in directing our steps toward the very thing we were so greatly in want of, we commenced at once to take off her skin and cut her into four, as we had a short distance to go to the place where we intended to take up our quarters for the night. We had a hearty supper that night. I think I mentioned before that the flesh of the cow-seal is very good, but apart from that we had good reason to know that "hunger is the best sauce." We were not long without fresh sorrow, however, for one of the two that Holding and I persuaded to come with us died the day after.

The three of us who remained had lived on the east end of the island for about a fortnight, where we were getting plenty of shell-fish, but very ill sheltered, with the wind off the water, to which we had our camp very near, when it was proposed that one of us should go back and see if any more of the other party would come to us. The man who turned with Holding and I said that he would go. So giving him part of what we had, he started to go to them, and to give them all the particulars of how we were getting on. We waited anxiously day after day in hopes of some of them joining us, but none came. At length our seal got all done; we were always on the lookout to see and get another but never had the chance of getting one, although we sometimes saw some in the water. We were obliged to take to our old method of living on roots and limpets. It was always roots and limpets, limpets and roots, day after day, but we had to rest contented with what we could get, although our hunger was never satisfied.

The other party never joined us and we began to think that there was something not right with them, or that the man who had started to go to them had lost himself in the bush and never reached them. We were getting very uneasy about them, so Holding said if I would stay where I was until he would come back, he would go and see what they were doing. He started accordingly, and returned the day after with the melancholy news that they were all dead except the captain and the second mate. The man that left us to go to them and the carpenter died the night after Holding joined them. The captain and the carpenter were both away gathering limpets, when the carpenter failed on the way home and died. We were now reduced to four in all. Holding said that the captain and second mate would join us soon, but that at present the second mate was unable to walk, as he had a very bad boil on his leg; the captain was to stay with him until he was better.

About this time we were getting some small fish among the rocks, for which we were very thankful. We also got some seaweed which we could eat. We shifted our quarters nearer to the place where we were getting the fish and the seaweed. It was of little consequence to us where we lived, as long as we could get food, and at the same time have a good view of the sea, which we still watched anxiously, in hopes of seeing some vessel that would rescue us from our miserable condition.

After a time the captain joined us, and said the second mate, who was not yet recovered, was to follow him as soon as his leg got a little better.

Up to this time we had never got another seal, and were still living on our old fare. We saw some seals at times, but not withstanding our best efforts, could not manage to get any killed. At last we one day caught a fine cow-seal after some trouble. We discovered her when we were on the rocks gathering shell-fish; she was asleep on a rock which was surrounded by water, but fortunately the water was not very deep, so that we managed to get outside of the rock on which she was sleeping. With our clubs we struck her several blows and killed her. Then we got her dragged up above high-water mark, and there we left her for the night, as it was beginning to get dark, so that we could not see to cut her, to enable us to carry her to our encampment. As well as I can recollect, this was about the end of the month of August, and the seals were very scarce at the time – hardly any were to be seen. We discovered an easier way of catching the fish, however, so that we could have generally some of them. In the way of fishing gear we had constructed a sort of basket out of reeds and strips of sealskin. With this basket or net, whichever it may be called, we caught a good many fish; but we could catch them only in moderate weather, and on the extreme east end of the island.

At this place we lived for a few months, with very little to eat at times; but at other times we could get more seals than we could use. They came by turns. Sometimes there would be a great many of them, and at other times we would not see one for some weeks. When the seals failed us we had to fall back upon our old fare, the limpets, fish and roots; but we were thankful that at all times we had a good supply of water.

We were always very careful of the seals' skins, which we stretched and used for a good many purposes, principally for making shoes. We had managed to kill some bull-seals, the skins of which we took very particular care of, as they were strong. Our intention was to build some sort of a boat, and we believed they would, in that case, become very useful to us.

We at last got the boat planned out, and commenced. We built her some two miles from the place where we lived, which distance we walked every night and morning. The reason why we built her so far away from our encampment was that we could not get the reeds nearer of which we made the framework. The reader will no doubt wonder what sort of model she was. I can assure him we thought her a very fine one. She was sharp at both ends; her bottom and sides were wrought in the same style as a basket. It required five of the skins sewed together with strips of themselves to cover the wicker work. We cut the skins into the shape of the frame, and then stretched them taut to the gunwales. She had three thauts [sic] or seats in her, and it may be believed she had no great weight when dry, but after being in the water for a short time, she got very heavy, as the skins then got soft and saturated with water. We had always to take her out of the water as soon as we had done with her. She was very leaky, as may be supposed, so that it always required one of us to be bailing constantly in order to keep her from sinking, and the worst of it was that she could only carry two of us at a time with safety, so that we could not do much with her in the way of looking after seals &c. She was on the whole a great trouble to us, but nevertheless she saved us many a weary march along the coast, among the scrub and rocks, as we were always on the look-out and moving about to see what we could pick up.

This was about the time that the young sea gulls were on the island. We got a few of them, and two or three eggs, which were very much relished indeed. We also got a few crabs now and then, but they were not so good as those got at home, nor anything near it.

We had a deal of trouble in keeping our boat in such repair as to be seaworthy she was always wanting something done to her before we went out, so that some of us were constantly employed upon her. She was unsatisfactory in the water, and when we drew her up upon the beach the sharp stones cut the bottom, so that we had always to be patching it up.

I may here mention that, before we made the boat, Holding proposed that he would go up and see what the second mate was doing. We were getting very anxious about him, as the Captain said, when he joined us, that he was to follow him soon. But, alas, when Holding went to the place he found him dead. It appeared that he had lived only a day or two after the Captain had left him, as the sort of almanac which we always kept, one of which was left beside him, was only marked up for one day after he was left alone.

The prospects for those who remained of us were very dismal at this time. We were now reduced to three, and the thought very naturally arose in our minds, "Who will be the next to be numbered with the dead?" We had now very little hope of any ship coming to relieve us from our desolate and miserable conditions.

We lived for some months at the place where we then were, getting a seal or two at times, and sometimes we had more than we could eat, as the flesh would not keep with us above eight or nine days.

Amongst our smaller discomforts, we were very much troubled with sand flies, which were very numerous in fine weather. There was also a sort of blue fly very abundant at the same time. We had a great deal of trouble in keeping the flies from the little meat that we had; we hung it up on trees as high as we could get, but still it was always fly-blown. It was the same with the fish, but we seldom had so many of them that they would be kept until they would be spoiled. The flies were indeed a great annoyance to us all the time we were on this island, but more especially during the summer months.

During the summer we had some days of fine weather, but with occasional showers of rain and heavy gales. When it did commence to blow, the gale generally continued for several days. But we were never idle. When it was blowing hard we were always busy repairing our boat. I think we required in all about twelve or fourteen skins for repairs during the time we had her in use.

It was now proposed that we should go and knock down the last house we were in which was built of wood, and try to build another and better boat with it. It was hard wood, and the boards were inch and half-inch thick. Two of us went and pulled the house down, and made a raft of the best pieces of the wood and brought it near to the place where we were camped. We then commenced with a hearty good-will to try our skill in boat-building; and I may say that we succeeded beyond our expectations. Very fortunately for us, the boards happened to be in three different lengths, which saved us a great deal of trouble in cutting them, as we had only the old worn-out adze and a few rusty nails, which we saved from the wood of the house. We had a good deal of trouble in making her water-tight, as we had little or nothing to caulk her with, except some old rags, which were of no use to us as clothes. All things considered, she answered very well. Soon after building her we went and collected all the wood left of the house, and rafted it near to the place where we were living, as we intended to build a sort of house with it, and make ourselves as comfortable as the circumstances would permit. But in this we were disappointed, for during the night a heavy gale came on which brought a great swell on to the beach, and carried off our boat before morning. We had no other help left now, but to commence and build another boat with the remaining wood. This renewed attempt was on the whole very successful. Indeed the boat was somewhat better than the other one – we had gathered some experience by our former work. I think this was about November or December.

The seals were now very scarce at the place where we were living and we thought that they had perhaps gone to some of the other islands. We proposed to go to a small island [Rose Island] to the eastward of where we were in search of them. We went, and found a few seals on it. There were thousands of rabbits here, but I am sorry to say that we caught but a few of them, and were glad to take what remains the hawks left of those that they had killed. We tried all sorts of traps to catch them, but it was of no use. Here we occupied ourselves several days in cutting branches and planting a hedge in the form of two sides of a triangle each side about 250 feet in length, so that when any of the rabbits went into the angle, we followed them, and drove them to the corner, but they invariably jumped over our heads, or over the top of the hedge, or ran through among our feet. I have seen as many as fourteen or fifteen in a time, yet we failed to catch any. We then tried to shoot them with a bow and arrow, but in this we failed also. They were too swift for us. We got, fortunately for us, some fish, but only at certain times, as there were very few sheltered places among the rocks on this island, and we had very often strong gales, which caused a heavy sea to beat on the rocks. Instead of fish, we managed to catch great quantities of widgeons at times. They were very tame, for we caught them when they were roosted on the rocks with a large stick and a running noose at the end of it. They were very good food when roasted, and, when the seals were scarce, they were our only support. On this island there were no wild roots, such as we had on the island we had left, and we missed them very much.

We had by this time nearly given up all hopes of ever getting away from these miserable islands, so we commenced to make a comfortable house for ourselves to winter in. We worked at it with a will, sometimes faintly hoping that some ship might after all come to our rescue.

The house we built was about ten feet square; the walls were made of turf about four feet high. We thatched it with brushwood, and lined it inside with seal-skins. We made a fireplace of stones, which we had to carry from the rocks. When finished, we found it very comfortable, I can assure you. The bottoms of our beds were of seal's skins stretched upon a stretcher, and then we covered them with some withered grass, and for blankets we used seal-skins. Our clothes were by this time all worn out, and we had to repair our shoes every week, or make new ones. We were very thankful that we could get the sealskins for the purpose. We made ourselves long leggings of the skins, which were a great comfort to us, especially when we were going through the scrub, as we had many a weary march in search of seals, and whatever we could find. We sometimes got a piece of a rabbit or two which had been killed and partly eaten by the hawks, and we were often very thankful for the least thing we could pick up. On this island we had a good view of the sea to the eastward, but day after day passed, and no ship appeared; nothing before us but the most dismal prospects.

All the time we little thought there were other five sufferers near us. I mean Captain Musgrave and four of his men, who were also wrecked on these islands, but who got off by their own exertions. Having saved some tools, and having the wreck quite handy, they built a dinghy, in which Captain Musgrave and two of his companions reached Stewart Island and dispatched a ship for the other two. We were deprived of all the means which they had at their command, so we could only trust in Providence bringing us relief in his own good time.

After we had been a month or two on this island the seals became very scarce. When we got on to it we thought we should fare very well, seeing so many tracks of seals, and so many wild rabbits and widgeons. But we had not been there long, before the seals were all gone, and, at times, we got but few widgeons, and the fish were also scarce. We had very hard work sometimes to keep the life in at all; yet, thank God, we always got a little of something.

It was about the month of March, I think, when we commenced to clear a small space of ground on which we intended to raise a few turnips and roots, the seeds of which we collected on the island that we had left.

On fine days our boat was of great service to us, as we could go with her from island to island. We had to be very careful, however, as there was a strong current in the channel, at the flowing and ebbing of the tides. On this small island we lived about four months, and for all there being thousands of rabbits upon it, we only succeeded in catching a very few; although, as I said before, we tried all sorts of traps and snares, and we were always very glad to take what the hawks chose to leave us.

One fine day, I think about the beginning of May, we went to a small island to the eastward of the one on which we were living, and found that there were apparently a great number of seals there. We killed two, and brought them along with us to the island we lived on. We thought we should remove to this island, as the seals were apparently so plentiful. It was no use staying where we could get very little to eat, when by shifting we could get plenty. The winter was just coming on, and it had indeed a very gloomy appearance. We had severe gales, accompanied by rain, yet never a day passed but we were out in search of food. It was on the N.E. side of the island where we generally found most widgeons, and at the distance of about a mile and a-half [sic] or thereby, from where we lived. Day after day, we went there in spite of the piercing cold and rain, in search of the widgeons as they were the only means of support that we had had for some time. I mentioned before that there were no wild roots upon this island, and we could find but few shell-fish. To increase our miseries, it would sometimes blow continuously for five or six days, so that there was no possibility of our getting out with the boat in search of food, and I am sure that we would have all died in a short time. It would have been madness to have attempted to go to the small island, where we should have likely got plenty of seals, in the small boat, and such a rough sea as we now almost always had.

But our dismal sojourn was near its end. Truly may it be said that "Man's extremity is God's opportunity."

After being on the Islands for the space of twelve months and ten days, we were rescued by the Spanish ship, "Julian", which, to our great and unspeakable joy, hove in sight on the 21st of May, 1865, came close in to the island, and sent a boat on shore, expecting to be able to get some relief, as she was leaky. She was from China, bound for Callao, with coolies. There was a plague raging on board, of which a great number of the Chinamen died. It was beginning to get dark before the boat landed, and the boat's crew had to stay on the island all night. We made them as comfortable as we could; we fried some seal which we had left, which some of them liked very well. We were very glad to find that some of them could speak English. Fortunately, next morning was moderate, so we launched their boat, and got off to their ship before dark, where we were very kindly received; we got a suit of clothes each, and were made extremely comfortable on our passage to Callao, where we arrived on 26th June, and so ends my simple, unadorned narrative of a memorable year in my life.

BIBLIOGRAPHY

Abbott, Elizabeth. ed. *Chronicle of Canada*. Vancouver: Raincoast Books (dist.), 1990.

Austin, K.A. *Port Philip Bay Sketchbook*. Adelaide: Rigby, 1970.

Baker, John Holland. *A Surveyor in New Zealand, 1857-1896. The Recollections of John Holland Baker*. ed. Baker, E. Whitcombe & Tombs, 1932.

Barton, G.B. *Remarkable Wrecks*. 1895. n.p.

Canadian Encyclopaedia, 2nd ed. Vol. II. Edmonton: Hurtig Publishers. 1988.

Carnegie, Hazel. *Harnessing The Wind*, Centre for Scottish Studies, University of Aberdeen, 1984.

Carrick R. *New Zealand's Lone Lands: being Brief Notes of a Visit to the Outlying Islands of the Colony*. Wellington: Didsbury, 1892.

Crichton, Vincent. *Pioneering in Northern Ontario*. Belleville, Ont.: Mika, 1975.

D'Urville, Durmont. *The Voyage of the Astrolabe – 1840. – an English rendering of the journals of Durmont D'Urville*. Ed. Wright, Olive, Wellington: A.H.& A.W. Reed, 1955.

Doxford Harris, Helen. *Digging for Gold*. Nunawading, Victoria. 1988.

Eden, Allan W. *Islands of Despair*. London: Andrew Melrose, 1955.

Encyclopaedia Britannica, 9th ed. Vol. XXI. New York: Henry G. Allen & Co. 1898

Escott-Inman, H. *The Castaways of Disappointment Island*. Christchurch, Caper Press Ltd. 1980 (reprint from London: Patridge, 1911).

Eunson, Keith. *The Wreck of the General Grant*. Wellington: A.H. & A.W. Reed, 1974.

Fraser, Conon. *Beyond the Roaring Forties, New Zealand's Subantarctic Islands*. Wellington: Government Printing Office Publishing, 1986.

Hartland, John. *Seamanship in the Age of Sail*. London: Conway Maritime Press Ltd. 1985.

Ingram, Chas.; Wheatly, P. Owen. *Shipwrecks, New Zealand Disasters, 1795-1936*. Dunedin: Dunedin Book Publishing Assoc. Edinburgh House, 1936.

Jeffery, W. *Century of Our Sea History – South Seas*. n.p. 1901.

Kerr, I.S. *Campbell Island: a History*. Wellington: A.H. & A.W. Reed, 1976.

King, M. *Moriori, A People Rediscovered*. Auckland: Viking, 1989.

Landstrom, Bjorn. *The Ship, an Illustrated History*. Garden City, New York: Doubleday & Co. 1961.

Lubbock, Bard. *Last of the Windjammers*, Glasgow: Brown, Son & Ferguson, 1927.

McLaren, Fergus. *The Eventful Story of the Auckland Islands*. Wellington: A.H. & A.W. Reed, 1948.

Malone, R.E., *Three Years' Cruise in the Australasian Colonies*. London: Richard Bentley, 1854.

Mariners' Handbook. Ed. International Correspondence Schools: Chicago, Philadelphia, Toronto: John C. Winston Co. 1921.

Morrell, B., *A Narrative of Four Voyages, 1822-1831*. New York: Harper, 1832.

Musgrave, Thomas. *Castaway on the Auckland Isles: a narrative of the wreck of the "Grafton" from the private journals of Captain Thomas Musgrave*, edited by J.J. Shillinglaw. (Melbourne edition published by H.T. Dwight, 1865.)

New Zealand Pilot, 12th edition, London: Hydrographic Department, Admiralty, 1958.

New Zealand's Subantarctic Islands, A Guide Book. Department of Conservation, Wellington: 1991.

Noble, Capt. John. *Port Philip Panorama: A Maritime History*, Melbourne: Hawthorn Press, 1975.

Phillips-Brit, Douglas. *A History of Seamanship*. New York: Doubleday & Co. Inc. 1971.

Raynal, F.E. *Wrecked on a Reef: or, Twenty Months among the Auckland Isles: a True Story*. (translated from the French) London: T. Nelson & Sons, 1874.

Smith, Andrew. *The Castaways: A Narrative of the Wreck and sufferings of the Officers and Crew of the Ship Invercauld of Aberdeen, on the Auckland Islands*. Glasgow: Brown & Son and Ferguson, 1866.

Sorensen, J.H. *Wild Life in the Subantarctic*, Christchurch: Whitcombe & Tombs, 1951.

Villiers, Capt. A. *Men, Ships and the Sea*. Washington: National Geographic Society, 1973.

Wakelin, R. *History and Politics: Leaves from the Writings of a New Zealand Journalist*. Wellington, 1877.

World Almanac and Book of Facts, Ed. Hoffman, M. S., Pharos Books, New York, 1992.

Scientific Papers, Abstracts and Reports

Alpin, P. *Possible Sites for a Permanent Base at Port Ross, Auckland Islands*. Feb. 1982.

Canadian Mines Register. Northern Miner Press, Toronto, 1935.

Falla, R.A. *A Vanished Township: "Hardwicke" or the Enderby Settlement*. 1971.

Falla, R.A. *Comments on the Enderby Settlement and the Cemetery at Port Ross, Auckland Islands*. 1971.

Dingwall, P. *The Changing Image of the Auckland Islands*. Landscape 9, Wellington, Department of Lands and Survey, May, 1981

Hurst, M. *Erebus Cove Archaeological Site Survey, Auckland Islands*. February-March, 1982.

Journal of the Polynesian Society, Vol. 2, 1893.

Management Plan for the Auckland Islands Nature Reserve, Department of Lands and Survey, Wellington, 1987.

Management Plan for the Auckland Islands Nature Reserve, (Draft) Department of Lands and Survey, Wellington, 1993.

New Zealand's Subantarctic Islands, Conservation Management Strategy. Department of Lands and Survey, Wellington, unpublished draft, 1993.

Report of the Bureau of Mines, Vol. 9, (1900), Vol. 22, (1913), Vol. 30, (1921). Government of Ontario, Canada

Ritchie N. *Comments on Aspects of Archaeological and Historic Site Management in the Auckland Islands*. March 1986.

Thorpe A.R. *Request for Plaques to Restore Graves on Auckland and Enderby Islands*. 1973.

Thorpe, A.R. *Report on Historic Sites and Existing Buildings visited during the Auckland Island Expedition, 1972-73.*

Transactions and Proceedings of the New Zealand Institute, Vol. II, 1869. Wellington: Government Printing Office, 1870.

Transactions and Proceedings of the New Zealand Institute, Vol. XXII, 1890. Wellington: Government Printing Office, 1891.

Yaldwyn J.C. ed. *Preliminary Results of the Auckland Islands Expedition, 1972-73*. Wellington: Department of Lands and Survey, 1975

Records, Reports, etc.

Armstrong, H. 'Official Report on Cruise of the Brig Amherst'. *New Zealand Government Gazette, Province of Southland*. Vol. 6, No. 9, 11 April, 1868.

Clifton, L. Auckland Islands. Unpublished report, Wellington, 1946.

Enderby, C. *The Auckland Islands: a short account of their climate, soils and productions; and the advantages of establishing there a settlement at Port Ross for carrying on the southern whale fisheries*. London, Richardson, 1849.

Norman, W.H., and Musgrave, T. *Journals of the Voyage and Proceedings of H.M.C.S. Victoria in Search of Shipwrecked People at the Auckland and Other Islands*. Melbourne, Victoria Government Printer, 1866.

Newspapers, Magazines and Periodicals

Aberdeen Herald. July 29, 1864 (Announcement of wreck); *Aberdeen Journal*. November 4, 1863 (Launch of the Invercauld); August 2, 1865 (Wreck of the Invercauld); August 9, 1865 (Obituaries); August 16, 1865 (Dalgarno's letter); September 6, 1865 (Smith's return); June 4, 1868 (Obituary of R. Connon); September 30, 1874 (Launch of Invercauld II); January 29, 1906 (Obituary of R. Connon, Jr.).

The Press, Christchurch. March 19, 1891 (Search for the *Kakanui*).

Dog Watch, Vol. 10. Melbourne, 1953. Parr, W.G., "Wreck of the *Invercauld*".

Marine News, Fleet list No. 51, January. 1965. Somner, Graeme., "Richard Connon, Reid & Co." Aberdeen.

Forest and Bird. May, 1984. Eagle, Audrey, "The Auckland Islands Visited".

Land, Sea and Air, Bateson, H. "Tragic Auckland Islands" 1921.

Leopard, Aberdeen, September, Short, A. "Wreck of the Invercauld." 1982.

Melbourne Illustrated Post, 25 Oct. 1865. (Account of wreck.)

Melbourne Argus, 13 October, 1865 (accounts of wreck) – source: *People's Journal* (Scottish Periodical).

New Zealand Geographic, Oct.-Dec., 1990. Pope. L. "Auckland Islands – Wild Splendour".

Pacific Way, MacDonald, B., "Realm of the Albatross", n.d.

Miscellaneous

Vaughn, Rose (Rose Vaughn Trio), "Stone and Sand" from the CD *Fire in the Snow*. Canadian Broadcasting Corporation, Halifax, Nova Scotia, Canada. 1993.

No Place For People, South Coast Productions, Riverton, New Zealand, 1996. Video documentary on the geography, wildlife, and history of the Auckland Islands. Running time: 45 minutes.

INDEX